THE BLIND SPOT

THE BLIND SPOT

Science and the Crisis of Uncertainty

∾

WILLIAM BYERS

PRINCETON UNIVERSITY PRESS

PRINCETON AND OXFORD

Published by Princeton University Press, 41 William Street,
Princeton, New Jersey 08540
In the United Kingdom: Princeton University Press, 6 Oxford Street,
Woodstock, Oxfordshire OX20 1TW
press.princeton.edu

Library of Congress Cataloging-in-Publication Data

Byers, William.
The blind spot : science and the crisis of uncertainty / William Byers.
p.   cm.
Includes bibliographical references and index.
ISBN 978-0-691-14684-3 (hardcover : alk. paper)
1. Science—Social aspects. 2. Uncertainty (Information theory) I. Title.
Q175.5.B94   2011
500—dc22
2010029682

British Library Cataloging-in-Publication Data is available

This book has been composed in Aldus

Printed on acid-free paper. ∞

Printed in the United States of America

1   3   5   7   9   10   8   6   4   2

Contents

Preface: The Revelation of Uncertainty

We are going to have to learn to live with a lot
more uncertainty for a lot longer than our
generation has ever experienced.
—Thomas L. Friedman[1]

This book is about science, what it is as opposed to what people say it is; what scientists do as opposed to what most people believe they do. Science is what we use to understand the world and to understand ourselves. It defines what is real and sets limits on what is possible, on what is conceivable. Today when the dream of unending and inevitable progress seems unduly optimistic, then perhaps the time is ripe to go back and reexamine our view of science and sort out its strengths from its weaknesses.

Most people would identify science with certainty. Certainty, they feel, is a state of affairs with no downside, so the most desirable situation would be one of absolute certainty. Scientific results and theories seem to promise such certainty. The popular belief in scientific certainty has two aspects: first, that a state of objective certainty exists and second, that scientific kinds of activities are the methods through which this state can be accessed. Yet I will make the case that absolute certainty is illusory and that the human need for certainty has often been abused with noxious consequences.

Contemporary society is beset by an ever-increasing set of crises and potential crises, which are exacerbated, and in some cases brought on, by a misreading of science and the scientific method, a misreading that we could call pseudo-science. This brings on a kind of vicious circle where the solutions that are proposed to the problems we face only succeed in making matters worse. Is there another, expanded, way of looking at science that will put the drive for certainty in perspective and provide a framework within which uncertainty can be seen as both inevitable and as an opportunity? That is the challenge of this book.

Many people have the naïve belief that every problem can be solved by science and technology, through the systematic application of certain practices and ways of thinking. Unfortunately, the leaders of government, commerce, and industry are among those who tend to hold these views. At the other extreme there are those religious fundamentalists who blame many of the ills of civilization on science and fantasize about some pre-scientific paradise. Both of these views elevate science and technology to the center of modern culture. In this respect, at least, both views agree and are correct. Science, both directly and through its influences, is the dominant element in modern civilization.

Yet modern civilization is in crisis! We face not just one crisis but a series of interconnected crises—the economic crisis, the environmental crisis, and the crisis in relations between the secular and religious worlds, especially the world of religious fundamentalism. There is a deep connection between these crises and the world of science and technology. In fact, a better way to think about the present situation is that what looks like a series of disparate crises is really one crisis that manifests itself in various ways—one all-encompassing crisis that arises from inner contradictions that are inherent in modern culture. The origin of this fundamental crisis is to be found in a misunderstanding of the nature of science. Unfortunately, such a misunderstanding is also quite common in those highly educated segments of society that are called upon to respond to the critical situation we face. One's response to any problem is constrained by one's understanding of it and so it is with our contemporary difficulties.

This book will demonstrate that our understanding of science is simplistic and thus inadequate to the task at hand. We must develop a more sophisticated understanding of what science is, and, as a consequence, what it can and cannot do for us. Most of us assume that science is monolithic—science is science is science—but this book will demonstrate that science carries within it diverse tendencies. To highlight these differences I give them different labels—the "science of certainty" versus the "science of wonder" but we should take the former to be a simplistic misinterpretation of the latter. What differentiates these two viewpoints is whether or not what I shall call the Blind Spot is acknowledged. The

Blind Spot refers to an intrinsic and inevitable limitation to scientific theories and even to scientific concepts.

All of the crises mentioned here can be traced back, in one way or another, to the point of view of the science of certainty. What Friedman said earlier about the economic situation seems to be generally true of our times—as societies and as individuals, we must learn to live with uncertainty. The two approaches to science that I discuss in this book divide up neatly in this regard; the first attempts to deny or eliminate uncertainty, the second takes uncertainty as an inevitable fact of life, as an opportunity, and considers how best to work with it.

Science and technology have come to define what is real, to define what is true. It is a well-worn cliché to say that in order to look for creative solutions to the problems we face, we must learn to "think outside the box." A certain ideology of science and technology constitutes the proverbial "box" in this instance, and we must get outside of it if we hope to deal with the present situation.

I emphasize that I am not condemning science and technology as a whole, nor am I ignorant of the many benefits that science has conferred upon the world. The problem lies not with science but with the point of view that I call the "science of certainty," a particular approach to science in which the need for certainty, power, and control are dominant. The identification of all of science with this particular tendency within science makes up a kind of "mythology of science." It is this mythology that is called upon when some governments, administrators, and businessmen misuse science to justify their questionable practices. This attitude has caused of a great deal of damage. It must be brought to consciousness if it is to be questioned and changed. It follows that in our search for what has gone wrong and what can be done to fix it, we must take another look at science.

It seems strange to call science a mythology since the story that science tells about itself is precisely that it is objective and empirical; that it concerns itself with the facts and nothing but the facts. This is often what we are referring to when we use the word "scientific." And yet science is a human activity, an activity pursued by human beings. This is an obvious statement but it bears repeating since part of the mythology of science is precisely that it is independent of human beings, independent of mind and intelligence. If these claims were correct then they would point to a mystery that is not always appreciated. How do human beings create a system of thought that produces results that are independent of human thought? For science certainly involves a particular way of using the mind. Most often we think of scientific thinking as rational thinking characterized by clarity and logic. Are these the only characteristics of scientific thinking? Are they even the most important ways in which scientists think? Can such thinking conceivably produce the grand creative leaps that science is famous for? Can such thinking help us escape from the "box" in which we find ourselves?

I mentioned that in these early years of the twenty-first century the dominant cultural force, for good and for bad, is science, and a certain way of using the mind that passes for systematic thought. Of course, the ways in which we think impose limits on the possibilities for change. We have a very restricted notion of how open-ended the situation is and so we feel that our possibilities are limited. We feel constricted by the scope of the problems we face and so we have almost lost hope.

But are the current problems that society faces really different in kind from the situation that every person who does creative work finds himself or herself in on a regular basis? Authentic scientific workers live in a world of change that brings with it the need to continually rethink the significance of what they are doing and why they are doing it. Science is an exercise in human creativity. As such, it must continually reach toward the unknown and the uncertain, toward the Blind Spot. It must look back to its mysterious and opaque sources in the human mind. We must learn to allow these creative sources to form part of the language we use when science is discussed and applied. We must find a new way to talk and think about science because the point of view that underlies most of our discussions is not only incomplete and therefore incorrect, it is dangerous!

The book is organized in the following way. Chapter 1 introduces the existence of an inevitable "blind spot" in our scientific theories, an unavoidable incompleteness in our description of reality. This blind spot arises out of human consciousness itself, and is rooted in the biology of the brain. In chapter 2, I discuss how specific breakthroughs in science and mathematics have revealed this blind spot. Modern scientific thought is permeated with the discovery of the uncertain in various guises. There are two possible reactions to these kinds of discoveries—one negative and one positive. The negative response involves anxiety and repression; the positive is connected to creativity.

Chapter 3 isolates and compares two different aspects of science that I called the science of certainty and the science of wonder. These contrast the need for certainty and rigor with the need for freedom and creativity. Science contains both of these tendencies, but they do not coexist comfortably—the two conflict. In chapter 4, this dichotomy within science is linked to some of the crises that the world is currently experiencing in the economic and political spheres.

I then go on to show how mathematics and science contain elements of the uncertain. Ambiguity can be thought of as a form of uncertainty, so chapter 5 introduces this central notion as a single idea that can be seen from conflicting points of view. Ambiguity is seen to be unexpectedly present in much of mathematics and science. In chapter 6, another kind of ambiguity is discussed, this time in the fact that the scientist is inevitably both participant and observer in his scientific work. This introduces the element of self-reference, observing oneself as a participant, which is central to my description of scientific activity.

Chapters 7 through 9 take up the task of finding a way to talk about a scientific world replete with uncertainty and self-reference. Chapter 7 begins a discussion of scientific concepts by considering the notion of number. Actually, it is claimed that "number" is not a concept but a proto-concept, by which I mean an idea that generates concepts. Number is seen to involve the related ideas of quantity and measurement but number also has qualitative properties that are usually not mentioned in a discussion of science. The distinction between quality and quantity lets us look at the whole scientific enterprise from an unusual perspective.

Much of the previous discussion has brought to the fore conflicting perspectives that are present in science and in the scientist. Chapter 8 sees these conflicts from the perspective of a deeper unity that forms the essential context for any deep discussion of science. Science is motivated by a desire to unify our experience of the world and to unify our selves with the world. Yet this unity that is discussed is a subtle affair; it is a divided unity. It is this division within unity that generates the complexity we find in our description of the natural world.

In chapter 9, I demonstrate that it is the self-referential element within science that generates its dynamism. Science that is alive and growing is a science that has a complex, fractal nature. This complexity has its roots in a description of reality in which ambiguity is fundamental and objective clarity is but one aspect of a deeper ambiguity. Finally, in chapter 10, I consider the lessons that this view of science can teach us, and how it might affect our approach to the series of crises that began the book.

THE BLIND SPOT

I

∞

The Blind Spot

There are, indeed, things that cannot be put into words.
They make themselves manifest.
—Ludwig Wittgenstein[1]

Ludwig Wittgenstein was one of the great philosophers of our time, and yet the preceding statement is among his more obscure, especially when thought of in relation to science. In this case, he is saying that certain aspects of science, though real, cannot be put into words. Einstein understood this very well when he talked about his feeling that "behind anything that can be experienced there is something that the mind cannot grasp and whose beauty and sublimity reaches us only indirectly and as a feeble reflection."[2] The "blind spot" is my name for those things that are real but which the mind cannot grasp and thus cannot capture through words, symbols, or equations. I will now give some inkling of what Wittgenstein and Einstein were talking about, even though trying to use words to indicate that there is something beyond words is obviously a strange, not to say paradoxical, thing to do.

Let me begin with an old joke. A drunken man has lost his house keys and is searching for them under a streetlight. A policeman approaches and asks what he is doing.

"Looking for my keys," he says. "I lost them over there." And he points down the street.

"So why are you looking for them here?"

"Because the light here is so much better," the man replies.

The "light" refers to language, concepts, and reason. There is, for example the expression "the light of reason." "Darkness" would then represent the reality that lies behind conceptual language, reality in its pristine form—precisely what science is attempting to investigate. When you think about it, this is a little strange. We are trying to describe the darkness, but we do it by turning on the light. Of course, when you turn on the light, the darkness inevitably disappears. Darkness is a metaphor for the blind spot and, for this reason, the blind spot does not refer to some particular fact that cannot be put into words or some specific situation that cannot be understood. The blind spot is implicit in every situation.

Think about young children before they have learned to talk. I have a granddaughter, Aviva, who has just turned one. She is a delight—so interested in exploring her environment, so excited by her new experiences, the new textures to touch, new tastes, and so on. The world for her is a world of wonder! The blind spot refers to this world of wonder. Of course, as Aviva grows up, the immediacy of the sensory world will recede as she acquires verbal and intellectual skills, but it will never disappear. It will always be there, ready to reveal itself to her in one of those magical moments that occur from time to time in everyone's life.

I have borrowed the expression "blind spot" from the psychological phenomenon of the "blind spot" in our visual field. The *physiological blind spot* is the place in the visual field that corresponds to the lack of light-detecting photoreceptor cells on the optic disc of the retina where the optic nerve passes through it. Since there are no cells to detect light on the optic disc, a part of the field of vision is not perceived. The brain fills in with surrounding detail and with information from the other eye, so the blind spot is not normally perceived.[3] It seems incredible that our visual perception is incomplete in this way; it goes against our inner conviction that the world we perceive is coherent and complete. But the existence of the visual blind spot is a good metaphor for the ungraspable element that we confront when we attempt to probe the natural world in our scientific work. Just as our brain provides us with the illusion that there is no visual blind spot, so our rational intelligence—through its insistence on consistency and completeness—hides the blind spot from our consciousness.

Another scientific metaphor for the blind spot is the phenomenon of the black hole, a "region of space-time from which nothing, not even light, can escape. A typical black hole is the result of the gravitational force becoming so strong that one would have to travel faster than light to escape its pull."[4] Because black holes exist but cannot be seen, they are a good way to think about the blind spot. Black holes contain singularities at their centers, points at which the equations of general relativity break down. These singularities are mysterious objects. Are they real or do they merely indicate a breakdown in a particular theory? Any physical theory that attempts to put together quantum mechanics and relativity will have to deal with the phenomena of black holes and singularities. I would argue that the relationship of black holes to a fundamental physical description of the world is analogous to the relationship of the blind spot to a fundamental philosophical description of science.

The experience of suddenly becoming aware of what was formerly a blind spot is shocking and disturbing. Consider the experience of the blind spot in your car. You decide to change lanes and so check your rearview mirror to make sure you have plenty of room to merge into the oncoming traffic. However, just as you start your move, a car you were not aware of, pops up, seemingly from nowhere. This is an experience every driver has had. It is disconcerting and a little embarrassing. Why? Because we realize with a shock that the mental picture we had of the cars on the highway is not identical to the actual situation. It takes a while to settle down again and regain confidence in the accuracy of our mental map.

The preceding metaphors and analogies have something to teach us about our scientific descriptions of the natural world. All such descriptions have inevitable spots that we are blind to precisely because it is the function of language and culture to hide them. Consider something that Stuart Kauffman, the theoretical biologist and complex systems researcher, said,

> My claim is not simply that we lack sufficient knowledge or wisdom to predict the future evolution of the biosphere, economy, or human culture. It is that these things are *inherently* beyond prediction. Not even the most powerful computer imaginable can make a compact description in advance of the regularities of these processes. There is no such description beforehand. Thus the very concept of a natural law is inadequate for much of reality.[5]

The statements of Wittgenstein and Kauffman contain the seeds of a different view of science, one that admits that there exists an intrinsic limitation to what can be known through science. It places science within a more open and spacious context and sets the stage for this chapter.

The existence of that which is real but cannot be understood poses a major challenge to our usual way of thinking about the world and to our thinking about the relationship between human beings and the natural world. Ask yourself if you believe that there are things that cannot, in principle, be understood. Your answer will tell you a great deal about yourself! The discovery of such "limits to reason" is in many ways the key scientific discovery of the twentieth century, one that our society has still not fully assimilated. I shall go into specifics in the next chapter but for now let me just say that it is this factor that explains the controversial nature of a good deal of modern mathematics and physics. I am thinking about Cantor's discovery of different orders of infinity; Gödel's proof that within any deductive system there are results that are true but cannot be proved; about the second law of thermodynamics that states the amount of disorder within a system must always increase; about uncertainty and complementarity in quantum mechanics; about the "butterfly effect" in the theory of chaotic systems, which says that every small change in the initial conditions of a system can have an enormous effect on its eventual state; and about randomness that seems to show up just about everywhere, from the theory of evolution to the fluctuations of the stock market. All of these point to intrinsic limitations in our ability to pin down reality in concepts and symbols. This is a key ingredient in the approach to science and mathematics that I am taking in this book, an approach that attempts to come to grips with the element of self-reference that is inevitably part of any attempt to describe the world as a living system.

ON DEFINITION

I think that there is such a thing as Quality, but as soon as you
try to define it, something goes haywire. You can't do it.
—Robert Pirsig[6]

What is a definition? In science, we usually think that a concept is captured by means of a definition. It makes the concept precise; it circumscribes the concept; it sets limits so we can now say precisely what is and what is not an instance of the concept. Such precision through definition is a necessary condition for a subject to be regarded as scientific. Without this kind of precision it would be difficult to imagine the process of measurement and quantification getting started. If I ask what a (mathematical) group is, the answer is a set whose elements can be multiplied in some reasonable way subject to some very specific requirements. The concept *is* its definition. Yet, as we shall see, many mathematical and scientific concepts point to something that is deeper, more all encompassing, than their definitions. Some things cannot be put into words because doing so is only an approximation to the real situation. The verbal or symbolic formulation captures some aspects of the situation but is not identical to it. There is a question regarding the relationship between the definition and the thing being defined.

To really grasp the essence of the problem with definition, one must go back to the Ancient Greeks. The Greek philosopher Parmenides is reputed to have maintained that you can only speak of what is, "what is not cannot be thought of and what cannot be thought of cannot be." It followed from this attitude that (absolute) "infinity" or even "zero" could not be defined because they "could not be." This attitude is a philosophical precursor to the "naïve realism" of today: the sense that the proper role of language is to enter into a one-to-one correspondence with the objects of the real world. It is a sensible reaction to the complexities of language. At first glance, it seems entirely reasonable to insist on a one-to-one correspondence between words and reality. Why? Because it protects us from the self-referential spiral that is inherent in human self-consciousness, the ambiguity that lies at the heart of the human condition that I shall discuss in subsequent chapters.

The problem of the relationship between language and reality is a problem that has been around for a long time. It is a vital problem, since without a clear notion of the nature of "definition" we cannot really begin to study mathematics or science. This question is a primordial one for any philosophy of science.

Is it always possible to keep the definition of a concept consistent with its meaning? Think about "zero." Zero represents "nothing," yet "zero" is not nothing—it is a digit, a number we use every day. The definition of "zero" is inconsistent with the meaning of zero. We can see why the Greeks could not entertain the idea of "zero," yet their math and science was the poorer for this omission.

Or take the concept of infinity. Infinity (in-finity) means non-finite, the essence of infinity is that it cannot be captured by the finite. Yet that is precisely what defining infinity does—it reduces infinity to something that is finite and manageable.[7] Again, the concept does not jibe with the meaning. This twist in infinity is the reason that infinity caused so much trouble historically[8] and remains a prime difficulty for students of mathematics. It was at the origin of the Greeks' attempt to distinguish between "potential infinity," which they felt was acceptable, and "absolute infinity," which they rejected. Potential infinity is essentially infinity as a process—for example, when you say that for every large number A and every small positive number ε there is an integer n such that $n\varepsilon > A$. Absolute infinity means treating an infinite collection as though it was one completed object, like when we treat an infinite decimal such as 0.121212 … as a single (real) number. The reason for the Greeks' rejection of absolute infinity was the one I gave earlier: any definition of absolute infinity would be inconsistent with the meaning of infinity.

This problem with infinity was at the origin of the controversy that arose concerning the work of Georg Cantor. He claimed to have defined (absolute) infinity, which most mathematicians of the time claimed could not be done *in principle*. Of course, what Cantor had done was reduce infinity into something that was *defined*, which was circumscribed and manageable. This is what it means to define something so it can be worked with mathematically or scientifically—something that one can understand, that has definite properties and is a potential source of theorems and examples. Nevertheless, the Greek problem with absolute infinity was not resolved by Cantor's definition, nor will it ever be by any other.

The difference between definition and meaning may be clarified by differentiating between what is called the denoted meaning and the connoted meaning. The definition *is* the denoted meaning, but the (larger) meaning includes the open-ended collection of possible connotations.

In *How Mathematicians Think*, I listed a whole series of mathematical concepts that contained variations of this twist. I called them "Great Ideas" because I feel that this twist (or ambiguity) has great value. Great ideas include things like randomness, zero, and irrational numbers. In a way, all great ideas cannot be pinned down definitively because they all contain variations of the problem that is present in the idea of infinity. Take randomness, for example. Gregory Chaitin wrote[9]

Borel's conclusion is that there can be no one definitive definition of randomness. You can't define an all-inclusive notion of randomness. Randomness is a slippery concept, there's something paradoxical about it, it's hard to grasp. It's all a matter of how much we want to demand. You have to decide on a cut-off, you have to say "enough," let's take *that* to be random.

It is not that randomness cannot be defined. On the contrary, defining randomness has great value; it is a huge creative accomplishment. Nevertheless, there remains an inevitable gap between the definition and what is being defined. Identifying the definition with what is being defined may cause us to lose touch with the openness, the incompleteness, of the original situation. The original situation, which may well have contained elements of ambiguity, and even of paradox, that gave birth to the definition, now disappears, to be replaced by a new situation with its own problems and creative possibilities.

One might imagine that the gap between the deeper meaning and the explicit definition, the sense in which mathematical situations cannot be defined or understood, is exceptional. However it is present in every formal mathematical structure that inevitably contains undefined terms. Here is what mathematician Marvin Jay Greenberg has to say about Euclidean geometry:

> … we cannot define every term that we use. In order to define one term we must use other terms … if we were not allowed to leave some terms undefined we would get involved in infinite regress.
>
> Euclid did attempt to define all geometric terms. He defined a "straight line" to be "that which lies evenly with all the points on itself." This definition is not very useful: to understand it you must already have an image of a line. So it is better to take "line" as an undefined term.[10]

Greenberg goes on to list undefined terms in plane Euclidean geometry: *point, line, lie on* (a point "lies on" a line), *between* (the point A lies "between" the points B and C), and *congruent*. This is the modern approach to developing an axiomatic system. Some basic ideas must remain undefined. We cannot pin down everything—there always remains a certain incompleteness.

Even ordinary mathematical ideas share, to a certain extent, this problem with definition and, as a result, we shall have to learn to think about definitions in a new way. Take, for example, the idea of "number." What is a number? It is scarcely possible to define "number"—it is so basic and elementary. The German mathematician, logician, and philosopher, Friedrich Ludwig Gottlob Frege, tried to show that the idea of number could be developed starting with the idea of "set." The idea was to establish a firm foundation for all of mathematics. His attempt ultimately failed because of the discovery of certain paradoxes that arise when one thinks of a set in a naïve way as just a collection of objects. But that is not the main problem with this kind of approach. The problem is that such reductionism causes us to lose touch with the very thing we are interested in understanding—here, the nature of "number," the deepest and most important source of mathematics. The fact that "number" can or cannot be developed from some other concept does not necessarily help us in our attempt to understand and explore "number." In a sense, number cannot be defined, and yet to leave it at that is somehow also dissatisfying. "Number" evokes a whole universe, an entire manner of looking at the world, which I shall discuss in some depth in chapter 7. This universe can only be explored, not captured. Every deep mathematical or scientific idea, like the idea of number,[11] evokes a whole world. Some of these situations have a consensual meaning—integers, rational numbers. Some, like real numbers, are more complicated. But mathematics contains many different kinds of numbers and there is no intrinsic limit to the capacity of mathematicians to produce new kinds of numbers in the future.

Trying to understand something often means trying to give it a definition, yet (as in the case of infinity or randomness) another definition is always possible. Each definition structures a certain field of mathematical or scientific thought. Certainly one definition may be better than another but even an excellent definition does not capture the informal domain out of which it emerges. It structures the informal situation. When we use the word definition, we are usually referring to this formalized version, and yet understanding a given situation necessitates the integration of both levels—the formal and the informal. You could say that it is impossible to understand the informal, but the formal situation also has its difficulties. "Understanding" demands placing something in a context. It implies having a "feel" for the situation in which the concept arises, not to mention the ability to use the concept in novel situations or solve problems not previously encountered. You cannot understand a definition by parsing it. You acquire an understanding by working with the definition in many different circumstances, by thinking about it, by solving problems involving the concept, and by making mistakes and learning from those mistakes. Understanding is a process without end. At a certain stage in the process, one can say, "I understand randomness." But in reality you can always understand it better, understand it differently. The better you understand it, the more grounded you are in the primal notion. Randomness is not a thing. In a way, it does not exist; it is open and inevitably incomplete. Yet every formal definition of randomness produces its own reality that needs to be understood.

All interesting and important concepts have definitions with this kind of depth. An explicit formulation is not *the* definition but should be thought of as an "entry point," the beginning of an exploration. We then work with this (tentative) definition trying to expand our understanding. We do this by exploring in two directions simultaneously—backward by evoking the informal situation out of which it arose, forward by exploring examples and consequences. In the process of this exploration, our understanding will be expanded and made subtler. This process may then be iterated a number of times. Each subject we explore should be thought of more as a "field" (like an energy field in physics) than a fixed and definite object. A field does not have a fixed objective meaning. It is much much larger than that.

THE UNGRASPABLE

The conclusion of the previous discussion is that, in the deepest and most profound sense, the things that make up the world cannot be defined, nor can they be understood or pinned down in any definitive way. This is the gap that has emerged in the order of things, a gap and a challenge that has the most profound implications for how we conceptualize the entire scientific enterprise. I'll refer to this gap by speaking of the ungraspable.

Science is a way of approaching the world; it consists of asking nature certain kinds of questions and of obtaining certain kinds of responses.

The entire world of science is grounded in human consciousness and rationality. In science, the world is described in a specific way, using a certain kind of language—and so reality is reduced to rationality. How accurate is the picture of reality obtained through science? The existence of the "ungraspable" implies that there are intrinsic limitations to the cultural project of reducing reality to rationality. In a manner that is paradoxical yet consistent with the lessons of scientific progress in the last century, I shall base my critique of science on recent developments in science itself.

Blindsight

The New York Times recently carried an article written by Benedict Carey about a man, T. N., who had been left blind by two successive strokes yet was able to successfully navigate a cluttered hallway full of potential obstacles. Brain scans showed that the patient had no visual activity in the brain's cortex—he was profoundly blind—yet he saw. How was that possible? "Scientists have long known that the brain digests what comes through the eyes using two sets of circuits. Cells in the retina project not only to the visual cortex—the destroyed regions in this man—but also to subcortical areas, which in T. N. were intact." Most people are not aware that they possess these alternative resources for processing visual information. In fact, Beatrice de Gelder, a neuroscientist at Harvard and Tilburg Universities and the researcher involved with this experiment, said, "The more educated people are, in my experience, the less likely they are to believe they have these resources that they are not aware of to avoid obstacles." The patient, a doctor, was dumbfounded that he could navigate the obstacle course.

I bring up this experiment because it has implications for our discussion of the ungraspable. To grasp something usually means to integrate it into our normal conscious rational view of things. The essence of what is going on here is that what has been eradicated is what you could call "conscious sight," the normal sight of the visual cortex. Because the visual cortex was destroyed, it became possible to bring subcortical faculties into conscious awareness and possibly restore some partial visual capacity to T. N. Perhaps we also normally think of science as though it were a function of only a certain part of the brain. We have other faculties that are at play in our interactions with the world but they are not normally accessible to our conscious self and therefore often do not show up on our scientific radar screen.

GUT FEELINGS

The next example comes from some studies in 1997 led by Antoine Bechara and Antonio Damasio as described in the book, *The Mind and the Brain.*[12]

> Volunteers play[ed] a sort of gambling game using four decks of cards and $2,000 in play money. All the cards in the first and second decks brought either a large payoff or a large loss ... cards in decks 3 and 4 produced ... small risk, small reward. But the decks were stacked: the cards in decks 3 and 4 yielded, on balance, a positive payoff. That is, players who chose from decks 3 and 4 would, over time, come out ahead.... A player who chose from the first two decks more than the second two would lose his (virtual) shirt.
>
> Normal volunteers start the game by sampling from each of the four decks. After playing for a while, they began to generate what are called anticipatory skin conductance responses when they are about to select a card from the losing decks. This skin response occurred even when the player could not verbalize why decks 1 and 2 made him nervous. Patients with damage to the inferior prefrontal cortex, however, played the game differently. They neither generated skin conductance response in anticipation of drawing from the risky decks, nor shied away from these decks.
>
> Bechara and Damasio suggest that, since normal volunteers avoided the bad decks even before they had conceptualized the reason but after their skin response showed anxiety about those decks, something in the brain was acting as a sort of intuition generator. Remarkably, the normal players who were never able to figure out, or at least articulate, why two of the decks were chronic losers still began to avoid them. *Intuition, or gut feeling, turned out to be a more dependable guide than reason.* [italics added] It was also more potent than reason: half the subjects with damage to the

inferior prefrontal cortex eventually figured out why, in the long run, decks 1 and 2 led to net losses and 3 and 4 to net wins. Even so, amazingly, they kept choosing from the bad decks.

The point of this story is again that people have capacities that they cannot bring into everyday verbal consciousness. Even though we can talk about these "gut-feelings," this does not really mean we are "grasping" the intuitive sense. Bringing intuition into consciousness means translating it into another mode of awareness and so changes the original "gut-feeling" into something totally different. It even involves shifting from one region of the brain to another. In this sense, our "gut-feelings" are ungraspable.[13]

The same phenomenon often occurs to me when I write. I may have read some article or had a discussion and I have a gut feeling that there is something there that is relevant to what I am writing about. At this stage, I'm not sure what the relevance is exactly but I begin to write it down and integrate it into the chapter I am working on. More often than not I eventually get something coherent down on the page. When I reread it, I may say to myself, "Ah! That's what I was trying to say." Yet at the stage of the inarticulate gut feeling, my rational self does not understand what I want to say; what I shall end up saying exists but has not yet been grasped.

"Stunned by What Is but What Cannot Be Put into Words"

Now the previous two sections can be taken in various ways. One could conclude that everything can potentially be integrated into rational consciousness and that this is the definitive mode of being in the world that corresponds to the way things are. That is not what I am saying. My position is that *what* is understood cannot be definitively separated from the mental facilities through which we understand. Grasping some topic or situation refers to a particular way of interacting with it. Inevitably, aspects of the world cannot be grasped in principle. This ungraspable nature of things does not only refer to gut feelings or blindsight. It is also a feature of our normal scientific conceptual universe. The ungraspable refers to a quality of intrinsic incompleteness that is inevitably associated with the conceptual.

Many people will be surprised by the assertion that some things cannot be understood. These people inhabit what I will call the "culture of certainty," who imagine that science proceeds by totally mastering some particular aspect of reality before moving on to the next bit in the way an army conquers foreign terrain. The "army" in this case would be rationality itself. If some aspects of reality cannot really be grasped, then science never conquers any territory. It explores territory, but even a territory that has been well understood may yield additional surprises if it is approached from a novel point of view—everything can potentially be understood at a more profound level. What we understand at any given moment in time can be thought of as a two-dimensional surface. The third dimension in this metaphor consists of the depth that potentially can be brought to the situation.

The realization that things cannot be grasped may be seen as either a disaster or an opportunity. It is not necessarily a problem unless you make it into one. The opportunity it represents involves opening oneself up to a world that is alive and vital—a world of wonder. It has these characteristics *because* the world is not pinned down, stable, and totally predictable. Remember when I say that the world has this potential that I am simultaneously saying each one of us also has this potential. We are also evolving, alive, open, and vital! Without what I am calling the ungraspable, there is no awe or wonder and everything is smaller and less interesting.

This brings to mind a statement by the philosopher Abraham Heschel:

> What characterizes man is not only his ability to develop words and symbols, but also his being compelled to draw a distinction between what is utterable and the unutterable, to be stunned by what is but what cannot be put into words.[14]

Heschel's statement provides me with the opportunity of saying something about all of those aspects of our personal and professional lives that have the same inescapable incompleteness I have been discussing. Everything we have talked about in this chapter can be construed both negatively and positively. Negatively, we are putting limitations on what can be understood. Positively, we are saying that what we are referring to exists. We are stunned by "what is but cannot be put into words." To take a step in the direction of the unknown, to look at science and mathematics from this point of view, is to begin the process of healing the rupture within ourselves and within our culture that has caused and continues to cause so much damage.

Randomness exists, but what it is, its essence, cannot be grasped in words. The reason for this anomaly is that randomness is paradoxical—it has the same sort of gap we discerned in our discussion of zero or infinity. I tried to capture this paradox in the following parody of an Aristotelian syllogism:[15] "Mathematics is the study of pattern. Randomness is the absence of pattern. Mathematics studies randomness." But even concepts like number or continuity that are not paradoxical have an ineffable core. I shall pursue this more in chapter 7 when I talk about the difference between concepts and what I call proto-concepts. The ungraspable essence of things is what Heschel is pointing to. That which is real but inexpressible is not something vague or mystical; it is something that is immediate and simple. It is the ground out of which the concept arises. It is "nothing" but not zero; it is "infinity" but not infinite cardinal numbers; it is "time" but not the real variable "t" or even the fourth dimension of Einstein's space-time continuum.

CONCLUSION

Science derives from a source that is not accessible to science. At first glance, this statement seems so strange that one has to resist the temptation to reject it out of hand. Yet many people have had precisely the same intuition, even the same experience, in certain moments of creative insight. What creative worker in any field has not had the feeling that the sources of creativity are inaccessible to the conscious mind? It is the fear of blocking the creative source that accounts for the reluctance on many people's parts to analyze their own creative process. A mystery underlies the creative process, and this mystery is its very essence. Calling it a mystery does not mean the sources of insight do not exist or that they are "mystical" and thus unreal. Calling it a mystery means the sources are inaccessible to the everyday conceptual mind. When you look toward the unknown, which is what you do in scientific work, it is evident that one will see a kind of "blankness" or absence of structure. If the scientist can solve the problem by conventional means or by merely rearranging previously well-established elements, then it would not be necessary to descend into the unknown. The known must be exhausted before one is forced to confront the unknown. The sources of creativity are by definition unknown, inevitably outside of the present conceptual universe, since the conceptual universe is itself the result of acts of creativity.

The unknown is the matrix out of which creativity is born. The birth of creativity, the dawning of insight, is wonderful but unpredictable. One can work hard but that does not guarantee success. One can prepare for it but one cannot program it nor anticipate when or in what form it will eventually appear. Creativity has its origins neither in the natural world nor in the world of concepts—it involves much more than the mere shuffling of well-defined conceptual categories as a computer would do. Where do new concepts come from? If anything, concepts are the results of acts of creativity and not the other way around.

Clearly, a philosophy of science must begin with what is real. However, science is not identical to reality; science is a description of reality. The basic difference is what I meant by the difference between darkness and light at the beginning of this chapter. What we need to do is investigate the relationship between the description and the reality that stands behind it. The first thing that is necessary is to break the mistaken identification of science with reality. Of course, science is not arbitrary; it has a profound relationship with what it describes. Nevertheless, science is not to be equated with the real. This is a statement that is completely obvious yet bears repeating since it is necessary to differentiate between science and the mythology of science, between what science actually does and the story that is told about it. Just as the brain renders invisible the physiological blind spot and gives the illusion that the visual field is continuous and complete, so the mythology of science has the function of hiding from view the holes in the fields of consciousness and rationality. So, like the child viewing the emperor's new clothes, it is necessary to point out this blind spot.

I attempted to do this by talking about the "ungraspable," but the danger is that one thinks of the ungraspable as something divorced from reality. The "blind spot" I am talking about is an inevitable consequence of our rational consciousness. We are aware of it as a lack, but when we turn our conscious mind to it, it inevitably disappears. Yet we can infer the existence of this domain by making a small shift in the way we look at things. The development of science in the last century contains many instances of the discovery and the rediscovery of this phenomenon under a plethora of different guises—ambiguities, paradoxes, incompleteness, complementarity, randomness, and so on. Are they not all, in one way or another, blind spots? And what is the assumption of rationality, the assumption of logical consistency if not the mind's way of "filling in" the holes in our rational universe. When we look for this blind spot intellectually, it seems to disappear and so we must infer its existence in an indirect manner. Nevertheless, we *know* that our intellectual world of thought and reason is not all there is. Anyone who is engaged in creative work in the arts or sciences appreciates that creativity does not arise from reason alone. Its ultimate sources are ungraspable.

It always comes as a shock to realize that one's view of things is inadequate. As I mentioned in my discussion of the driver's blind spot, the realization that a blind spot exists can lead to a certain anxiety, the kind that people feel during an earthquake when you cannot rely on the stability of the earth beneath you. Yet when one attempts to plumb the depths of scientific thought to its very origins, one inevitably encounters this phenomenon that is extraordinarily difficult to explain or describe. It is immediately rejected by the rational mind, the point of view from which science is usually discussed. Thus, discussions of science and even the philosophy of science are usually after-the-fact rational reconstructions of science. Yet rationality is itself the result of an act of creativity and so cannot be used to explain the origins of the extraordinary creativity of science. Nevertheless, I maintain that without making this attempt to plumb the deep sources of scientific creativity we doom ourselves to a pallid superficial description of the scientific enterprise. We will miss the essence of science and the consequences will be grave, for it is our understanding of scientific culture that today determines what we think is real, not only in the natural world but also in ourselves. To put it bluntly, are we sophisticated machines or are we free open beings whose birthright is a kind of unlimited creativity? These are the stakes and I cannot stress too highly the importance of the view we hold of science, and as a consequence the view we have of what it means to be human.

2

≈

The Blind Spot Revealed

Ω can be interpreted pessimistically, as indicating there are
limits to human knowledge. The optimistic interpretation,
which I prefer, is that Ω shows that one cannot do mathematics mechanically and that intuition and creativity are essential.
Indeed, in a sense Ω is the crystallized, concentrated essence
of mathematical creativity.
—Gregory Chaitin[1]

INTRODUCTION

If the blind spot that I have been discussing in the previous chapter is
really present in all situations, then surely scientists must have encountered it on a regular basis in the pursuit of their scientific investigations.
This chapter will expand on the manner in which the blind spot can be
seen to underlie some of the most dramatic results of modern science. It
gives rise to the phenomena of ambiguity, uncertainty, and incompleteness that I will discuss in the following chapters. I will also discuss the
possible reactions to these encounters with the blind spot, with that aspect of things that cannot be grasped or understood.

One of the fascinating things about science is the extent to which it is
self-referential. The usual way of thinking about science is that it reveals
the deeper structure of the natural world. Sometimes, however, it also
says something about the process of scientific investigation itself. It is in
this element of self-reference that we will find the *meaning* of science.

Some people might dispute that science has meaning since for them it is like a photograph, an objective account of what is really out there. However, as we shall see in chapter 6, this view is just an approximation of what is really going on. You take this stance for as long as you can get away with it, but many indications suggest we cannot afford to indulge in such simplistic thinking at the present stage of scientific development. A more complete view is that there is no human endeavor without a context that supplies meaning. The interesting thing is that there are times when the content has implications for the context, when the theories of science tell you something important about the nature of science and thus its social and cultural role.

Consider the statement by mathematician and IBM research scientist, Gregory Chaitin, which began the chapter. Chaitin came up with an extraordinary real number, Ω (sometimes called the "Chaitin Number"), that cries out to be interpreted in a self-referential way. In Chaitin's own words, "Ω is the probability that a self-contained computer program whose bits are picked one by one by tossing a coin, will eventually stop, rather than continue calculating forever."

Surprisingly enough, the precise numerical value of Ω is irreducibly complex,[2] which means there is no way to simplify this number. Its decimal representation contains no patterns; there are no regularities to discern; it sits there impervious to our efforts to simplify it. And yet, it has great significance on a number of levels. It is a good example of mathematics' mysterious ability to capture aspects of things that are very profound.

As metamathematics, Ω has the two possible interpretations that Chaitin mentions. These point to the two different ways of looking at the entire scientific enterprise that I will discuss in the next chapter.

The "optimistic interpretation" highlights the creativity of science, while the "pessimistic interpretation" puts the stress on the limits to reason. Of course, these are just two reactions to the same situation and represent two possible reactions to the blind spot. Ω indicates the existence of a limit to total knowledge. This point of view feels negative, for it is a limitation on the scope and power of analytic intelligence. Yet, as we can infer from Chaitin's comments, on the other side of this negativity lies something of great value—the creative intelligence itself. Here we have all the themes of this chapter—the appearance of the

blind spot in the form of the number Ω and the two possible reactions to this discovery.

When I discuss our negative reactions to the existence of such restrictions to knowledge, I realize that the uncertainty of the unknowable is deeply unsettling to the analytic intelligence. We would like to say it is all a great mistake, that there is no such thing as the unknowable, yet so much evidence has piled up that it is difficult to ignore.

We can enter into this problem by considering our own emotional reaction to the previous discussion of the blind spot, which may well strike many readers as frustrating and possibly irritating. How could anything be unknowable? If there is something we don't know at present, how can we assert that we will not figure it out at some point in the future? Isn't every aspect of existence knowable *in principle* at least? For me, as a university professor and mathematician, the idea that there are limits to what can be known was initially deeply disturbing, and it still upsets some deeper parts of me.[3] I assume that most of you are similar to me in that regard; we are intelligent, equipped with well-developed analytic skills that we have cultivated over the years. These skills are what we rely on to help us navigate our way through the vicissitudes of life—these skills bring with them a remarkable degree of social status, power, and control. The well-trained analytic mind is beautiful; it is such a subtle, precise, and powerful instrument; no wonder so many of us are mesmerized by it. Now I am not saying that there is anything wrong with logic, analytic thinking, or the intellect, but they do have a tendency to take over, to insist on being the sole arbiter of reality. As a consequence, our reaction to the suggestion that our powerful, infinitely subtle, intellect cannot penetrate to the deepest layer of reality is initially to reject it out of hand. The whole thrust of our culture is to map out a rational universe using our rational minds. Science is the cutting edge of this tendency—it promises to make the unknown known, to clearly display the whole of the universe to our mind's eye. Unlimited understanding through science promises to bring in its wake unlimited power to control our environment, eradicate disease, and perhaps one day even defeat death itself.[4] This is the dream of reason—the dominant mythology on which our culture is based.

This dream misses something hidden by the very instrument, the light of reason, which we use in our search. After all, this entire con-

versation is happening at the intellectual level. To use words to point beyond words; to use well-formulated sentences to point out that there exist aspects of reality that are not captured by well-formulated sentences is a form of mental judo. It points to a different use of language than that with which we are familiar. Yet if we were talking about music, everyone would recognize the difference between the music itself and a description of the music. So it is with science and the natural world that science describes.

Reality is not a data bank, nor can it be reduced to a finite set of laws or to a deductive axiomatic system. Any system inevitably misses something. Words, symbols, and systems are all incomplete. Strangely enough, what is missed is not so much some extremely complex and subtle theoretical structure, a kind of superstring theory. The essence of what is missed is something that is not at all abstract or complex. It is the very process of abstraction, the representation of reality by language and mathematical symbols, which misses something.

Yet there is no intrinsic barrier between science and the natural world. The physicist Arthur Zajonc,[5] in speaking about the statement "Every new object, well contemplated, opens up a new organ of perception in us" (made by the great German man of letters, Johann Wolfgang von Goethe), said

> You have to live in that world of phenomena. You have to attend carefully every object well contemplated. Not just casually contemplated but well contemplated, attended to, over time, repeatedly. [This] changes who you are, changes who you are to the point that you begin to see things that you didn't see originally and perhaps that no one before you has seen.

"Attending carefully" is the doorway to wonder and a new view of science.

So what is missing from the conventional view of science? From the neurological perspective, what is missing could be called the view from the right hemisphere. Language and analytic thought are located in the left hemisphere so that any discussion of scientific theories is inevitably done from that particular perspective. How do we communicate to the part of ourselves that has no language, and how does the part of us that has no language communicate with that aspect of our intelligence that is bound to language and symbols? It is as though one part of us knows but cannot speak, while the other speaks very well but inevitably misses

that unmediated connection with things that is the content of the deepest form of knowing.

Another way of putting this is that what is missing from most scientific discussions is the role of the scientist as a subjective participant in the scientific enterprise. Discussions about science—scientific discourse itself—privilege the observer. From the point of view of the observer, the participant does not exist. One can talk objectively about the participant (as I am now doing), but the discussion is all on the home turf (so to speak) of the observer. Reason is grounded in one modality of human consciousness and thus is inevitably blind to other ways of knowing. Since the subjective aspect of things is independent of the objective and lies outside of it, there are limits on what we can know or grasp.

There is another way to encounter this domain, which is unknown from the point of view of the intellect, but, when encountered, is what I referred to earlier as a realm of mystery, wonder, or merely as light. This approach, which is accessible to anyone, is the way of introspection, or what Zajonc calls contemplation. This means to examine closely one's own mind, one's own thinking and creativity, with the intention of probing them to their source. Where do they come from? When you talk to a friend, where do the words come from? Do you prepare a thought and then speak it aloud? Of course not! The words come out spontaneously as a reaction to the situation. Where, then, do they come from? If you are honest, you may find yourself forced to say, "I don't know." However this feeling of "not knowing" comes in different varieties. It may be something negative, a lack, a source of frustration to be overcome. Or it may signify the positive, a condition of openness and freedom, the spontaneous expression of a basic intelligence.

THE BLIND SPOT AS IT APPEARS IN SCIENCE

Now the thesis of this chapter is that there have been a series of major scientific breakthroughs in the last century that all point in one way or another to limits on what we can know. These theoretical results are very strange. They tend to involve a certain degree of self-reference and so, as I mentioned earlier, they come very close to the paradoxical.

Modern science contains a new way of looking at reality, and what is distinctive about the new view is the emergence of these limits—limits to reason, to deductive systems, to certainty, and to objectivity. These kinds of discoveries have rarely been grouped together because the *scientific context* for them differs, but I maintain that they are all discovering variations of the same primordial situation. Whether the subject is set theory, mathematical logic, algorithmic randomness, chaos theory, quantum mechanics or statistical mechanics, the results carry a family resemblance. There are limits to what we can know.

I shall now proceed to sketch out some of these paradigm-changing scientific discoveries. I maintain that the scientific discovery of uncertainty is like a canary in a coal mine—it has been giving us a message but the message is so radical to our normal way of thinking that we have been unable to absorb it. At first glance, nothing about the scientific problems I shall discuss ties them directly to the social, economic, and environmental problems we face today. Yet the connection, I maintain, is there. It is in us, in our worldview.

Many of the examples that follow come from mathematics, so I must say a word about the relevance of mathematics to the present discussion of science. It is not just that mathematics is the language of much of science. Much of science involves quantification, which puts data in a form amenable to mathematical analysis. This analysis is often pursued using sophisticated mathematical models and techniques. But beyond all of this lies another way in which a discussion of mathematics is relevant to a description of science. Theories in science are often deductive logical systems. This formal way of describing particular theories has been studied more carefully in mathematics than anywhere else. Mathematics isolated this way of thinking millennia ago and has studied it ever since. Therefore, it is through mathematics that we can best study the strengths and weaknesses of this mode of thinking.

THE LOSS OF CERTAINTY IN MATHEMATICS

Mathematics is the traditional bastion of reason and absolute certainty. I have on my bookshelf an old book by the mathematician Morris Kline, *Mathematics: The Loss of Certainty*.[6] In the introduction he says, "It is now apparent that the concept of a universally accepted, infallible body of reasoning—the majestic mathematics of the 1800s and the pride of man—is a grand illusion. Uncertainty and doubt concerning the future of mathematics have replaced the certainties and complacency of the past." Kline was one of the thinkers who pointed out that something fundamental had changed in mathematics; the culture of certainty had been breached and things would never be the same again.

Uncertainty has entered mathematics in many different guises. An early shock was the creation of non-Euclidean geometries. For millennia, people had looked to Euclidean geometry as the quintessence of certain knowledge and it was the way that the mind was used in geometry— deductive logical reasoning—that seemingly brought elusive certainty within the grasp of humankind. Not only were the results of Euclidean geometry thought to be certain, they also seemed to reflect the very structure of physical reality. The deductive axiomatic method that Euclid introduced was *the* method for arriving at certainty both about the results themselves and about the properties of the natural world. Our culture has still not completely gotten over the blow represented by the emergence of non-Euclidean geometries.

The actual discussion revolved around the status of Euclid's parallel postulate, which states that if you are given any line, **L**, and a point, **P**, outside of that line, then you can draw a line through the point **P** that is parallel to **L**. Not only does this postulate seem reasonable, but given a piece of paper and a straight edge, anyone can actually draw the line in question. The parallel postulate seems to be *true* and so it would seem that it could be *proved* to be true, starting with a reasonable set of assumptions (axioms) such as those with which Euclid began his study of geometry. Non-Euclidean geometry stems from the belated realization—after millennia of effort by the best mathematical minds—that this postulate could not be proved from the other axioms. In a way, this is an early example of something that was *true* but could not be *proved*. This was an early encounter with the unknowable. Something had gone wrong with the previously infallible system of deductive thinking. What a shock this was!

Since the parallel postulate was independent of the other axioms, there were legitimate geometric systems in which this postulate did not hold, and each had an equal claim to validity. Geometrical truth was not absolute anymore, but depended on the geometric system in which you were working. For example, the (Euclidean) truth that the sum of the interior angles of a triangle is 180 degrees is replaced by a relative truth in which the sum is less than 180 degrees in the hyperbolic case and more than 180 degrees in the parabolic case. The existence of non-Euclidean geometries meant that geometry was not identical to reality anymore but had become a model of reality. Models are not true a priori (as the philosopher Immanuel Kant believed was true of Euclidean geometry), but need to be justified by their correspondence to the empirical data. As a result, certainty was never again as secure as it had been earlier. Even Euclid's system was discovered to have its faults. Though these were "corrected" by the great German mathematician David Hilbert, the climate had changed irrevocably—something, a quality of trust in absolute certainty, had been lost, never to be recovered in the old way.

As a result of this crisis, logic and the axiomatic system emerged as *the* way in which the equilibrium of certainty could be reestablished. Logic in the guise of a deductive system would, it was hoped, put mathematics and science back on a secure footing. These attempts were remarkably successful for a time. I mentioned Hilbert's "fix" for Euclidean geometry. Then there were successful axiomatic treatments of non-Euclidean geometries. Calculus presented quite a challenge. It worked and provided correct answers, but precisely why it worked was not clear. It took a hundred years to successfully provide secure foundations for the calculus and subject it to a fully logical development.

What is called the "arithmetization of calculus" was one of the great achievements of modern mathematics. It created the subject that is now called (mathematical) analysis that provided a secure foundation for calculus. Analysis was made more certain because it could be built up rigorously from a foundation of real numbers. Unfortunately, this gain in certainty came with a steep price—the emergence of another kind of uncertainty. This true cost of introducing the real number system only emerged later when deeper properties of the real numbers began to be recognized. This brings me again to the work of Georg Cantor.

Cantor and Infinity

It is Georg Cantor who deserves the credit for revealing some of the incredible complexity that was implicit in the real number system, which forms the context of almost all of the hugely successful applications of mathematics to science, business, and industry. Cantor was an extraordinarily original mathematician who did groundbreaking work on the nature of infinite sets. He has been the subject of innumerable studies because his work and his own personal history are so fascinating. Cantor developed a new theory of infinity that is today considered to be a basic part of mathematics. Yet Cantor's work evoked the hostility of his critics who actively tried to prevent him from publishing or receiving any academic recognition. In *Dangerous Knowledge*, a recent film produced by David Maloney for the BBC, the narrator quite correctly comments that, "Cantor frightened his critics. They saw mathematics as the pursuit of clarity and certainty. Everything Cantor was doing, his irrational numbers, and his illogical infinities seemed to them to be eating away at certainty.… It seemed Cantor had opened maths to the very thing it was supposed to save us from, irresolvable uncertainty."

Cantor is famous for introducing infinite numbers into mathematics in a systematic and rigorous way. In doing so, he dared to challenge a very powerful taboo regarding the use of infinity that had been in place for millennia. We owe him an enormous debt for his courage and tenacity, which not only provided us with some interesting mathematics but also opened the door to some fundamental insights into the nature of mathematics.

Some of the most far-reaching implications of Cantor's work involve the nature of the real number system. Most of us learned about the "(real) number line" in school. We learned to visualize all real numbers as points on this line: positive and negative integers, fractions, square roots, and transcendental numbers like π all have a specific location on the real line (see figure 1). In fact, the real line is such a pervasive picture for numbers that it is easy to forget it is not identical to the real number system but merely a model. We imagine that the real number line is *real* and that the various numbers all have a specific physical location on this line but, of course, the real line is not a physical object, it is a metaphor.

Because this geometric metaphor for the real numbers is so simple, we sometimes make the mistake of thinking that the real numbers are simple. The opposite is true—the real number system is one of the most complex and important creations of the human mind. It is basic to the development of calculus and differential equations, and so much more. Without the real numbers, there would be no science, as we know it. This

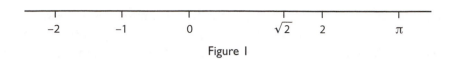

Figure 1

much is commonly understood but most people are not aware that our acceptance of the real numbers as our basic mathematical context means we have agreed to enter a strange and paradoxical world. It was Georg Cantor, more than anyone, who brought out the implications that are inherent on basing our mathematics on the real number system.

A real number can be thought of as a finite or an infinite decimal like 0.25 or 0.121212.... Every real number can be written as a decimal and every decimal represents a real number. The rational numbers have especially simple decimal representations: either repeating, as in the case of $1/3 = 0.333...$, or finite, as in $1/4 = 0.25$. Irrational numbers, like the square root of two or π, have decimal representations that are infinite and non-repeating. Thus, the decimal $\alpha = 0.12122122212222$ ... must represent an irrational number (the number of 2's keeps growing, so it can't be repeating). Now you might argue that a rational decimal like 0.25 or 0.121212 ... is a finite object since 0.25 is finite, and you could argue that the number 0.1212 ... could be described as "12 repeating" or as 12/99. An irrational number, on the other hand, contains an infinite amount of information. It is a complex, perhaps paradoxical, object. On the one hand, $\alpha = 0.121221222$ ... is one number with a precise value and location on the number line. On the other, it is an infinite object, the limit of an infinite sequence of rational approximations, in this case: $0.1 = 1/10$, $0.12 = 12/100$, $0.121 = 121/1000$, $0.1212 = 1212/10000$, and so on.

Is it reasonable to even say that all possible irrational numbers exist? And if they do exist, is their existence as secure as the existence of the numbers 1, 2, and 2/3. The mathematician Leopold Kronecker famously wrote that "God created the integers, all the rest is the work of Man." What he was saying is that the integers are discovered. They are independent of us, implicit in the natural world, but that the irrationals are invented. What is meant by the existence of the irrational numbers is just one of the problems associated with the real numbers that came to the fore because of Cantor's work.

Cantor forced the mathematical world to confront its ambivalent relationship with the idea of infinity, which was a problem for mathematicians and philosophers all the way back to Aristotle. The Greeks tried to contain this problem by carefully distinguishing between "potential infinity" and "actual infinity," where the former was considered to be acceptable but the latter was not. Cantor made systematic use of actual infinities and, as a result, his work is characterized by various results that seem paradoxical. These include, for example,

1. A whole, infinite set could be "equal" to a part of that set. Thus, the set A = {1,2,3,} has the "same" number of elements as B = {0,1,2,3,}, even though A clearly seems to have fewer elements.[7]
2. There are more real numbers than rational numbers, even though both collections are infinite.
3. One of Cantor's counterintuitive results was shown in the demonstration that "most" real numbers are transcendental (like π)—that is, they are not the solutions to any algebraic equation in the sense that the root of two is a solution to the equation $x^2 - 2 = 0$. It is notoriously difficult to prove that any *particular* real number, like π or e, is transcendental. "Almost all" real numbers are transcendental, but very few specific numbers are (can be shown to be) transcendental.

This final result was shocking and counterintuitive in itself, so much so that many people questioned the validity of a line of argument that could produce such a controversial result. To many mathematicians at the time, the argument, though logical, seemed more like magic than mathematics. What had gone wrong? Perhaps it was the form of reasoning that was suspect. In this kind of argument, the demand for logical consistency, the absence of contradiction, is used to demonstrate the ubiquity of a certain class of numbers—irrationals or transcendentals. Even though this kind of "proof by contradiction" had been around since the time of Euclid, there was a feeling that this particular usage went too far. If this kind of reasoning was truly illegitimate, then banning it might force mathematicians to redo most of mathematics. One school of mathematicians attempted to do just that, but what they produced was another variety of mathematics in the sense that non-Euclidean geometry is another form of geometry. The question about whether the method was wrong or the results were illegitimate was what could be called a generative question. Either way you look at it, Cantor's work forced mathematicians to see large parts of mathematics in a new light.

Today, given the general acceptance of Cantor's work, it is difficult to appreciate how problematic Cantor's mathematics once appeared. Chaitin commented in *Dangerous Knowledge*, that "the concepts that Cantor played with are intrinsically and inherently self-contradictory and people don't like to face up to that … he was playing on the edge. He had these ideas and you had to be very careful because at any moment they would bite you." Cantor paid the price for his ideas with a series of mental breakdowns that necessitated extended recuperative periods in an asylum. These breakdowns often came after intensive work on one of the most problematic of his conjectures—what is now called the continuum hypothesis.[8]

Cantor worked tenaciously on this problem, but it would not yield to his efforts. In fact, it was later shown[9] that the continuum hypothesis cannot be resolved in any conventional sense. It cannot be resolved in the same way that you cannot decide whether Euclid's parallel postulate is either true or false. Cantor's work, the hostile reception it received, coupled with his conviction that the work was vitally important (as well as the ambiguous nature of the problems he worked on), all of this threw him into a condition of extreme uncertainty. Science often has the effect of protecting the scientist from his own existential ambiguity, but in Cantor's case, his work resonated with that ambiguity and amplified it. His case reveals what is potentially at stake in pursuing certain kinds of scientific activities. Of course, this is only true if you take your science seriously and don't think of it as a mere game that has nothing to do with your life and the real world.[10] Cantor believed in the deep significance of the results that he was obtaining. As the author and mathematician Amir Aczel put it,[11] he felt that "infinity was the realm of God." He may well have felt that the nature of God was being revealed by his research into the nature of infinity—that he was accessing the realm of the numinous. If he felt he was actually making the unknowable tangible, then that partially explains why he paid such a heavy personal price for his creative breakthroughs.

Real numbers, it seemed, were objects that could only be "known" in a very limited way and not at all in the complete way that people believed they understood integers or rational numbers. The real numbers settled certain questions of the deepest mathematical and scientific importance, but these very successes opened the door to a whole series of seemingly paradoxical situations that threatened to destabilize mathematics. All of this contributed to a loss of faith in the objective and absolute certainty of mathematics.

This initial encounter with uncertainty had pushed mathematics into a stricter reliance on logic and accentuated the tendency to look at certainty as the defining feature of the subject. Logic had previously co-existed with mathematics, but because of the various crises I discussed earlier, it came to be seen as the defining characteristic of mathematics. This way of thinking about mathematics is called *formalism*. At its extreme, it became *logicism*, where mathematics has no meaning; there are no mathematical ideas, just logical and mechanical derivations from axioms, all of which can be described using finite sets of symbols. Not many people really believed the more extreme versions of formalism, but it became a common way to talk about the nature of mathematics. Nevertheless, formalism made a fundamental change in the nature of mathematics; it stripped mathematics of its *significance*, of what had been for centuries the very essence of mathematical thought. Mathematics began to seem like a kind of game—diverting, challenging, but divorced from reality. In this view, there is no truth, only a certainty that is obtained at the cost of excising the human elements from mathematics. With this approach, the wonder of mathematics is hidden from view and what remains visible is sterile certainty.

Gödel: Certainty Is Not Truth

The existence of fundamental limits to knowledge was revealed in a most startling and revolutionary way through the work of Kurt Gödel who was perhaps the greatest logician of all time. He definitively refuted the view that the truth of mathematics is captured by its formal deductive structure. One serious consequence involves the existence of a limit on what can be accomplished through the use of mechanically enhanced extensions of human rationality—that is, through the use of computers and computer models.

If we identify certainty with proof, then what Gödel had shown was that certainty is not identical to truth. He was a Platonist and believed in the truth and the objective reality of the mathematical world, its concepts and theorems. Attempts like that of Bertrand Russell and Alfred North Whitehead to create a foundation for mathematics that relied on logic were seen, in the light of Gödel, to have limited value.[12] The gap between certainty and truth had now been revealed for all to see. Gödel's result was not just a convincing opinion, it was a theorem with a proof. As a result, Gödel and the rest of us can be *certain there is no certainty!* Talk about a subtle, paradoxical, and self-referential result![13]

Gödel might have claimed that the loss of absolute certainty didn't bother him since he distinguished between certainty and truth. However, it is conceivable that Gödel's great achievements and his tortured personal life arose from the tension between truth and certainty. Gödel was a great logician and perhaps as a consequence he, like many mathematicians, scientists, and philosophers, leaned heavily, perhaps excessively, on his rational faculty as a guide to his personal and professional life. Reason appears to provide a safe harbor, an oasis of certainty, something on which we can depend in an uncertain world. Unfortunately, there inevitably are elements of one's life that cannot be controlled or proved. For example, one way to prove that your food has not been poisoned is to let your wife taste it first. If your wife is in the hospital and there is no one to prove that your food is safe, a logical response is not to eat at all. This is what Gödel did and, as a result, starved himself to death. Most people do not take logic to such an extreme but Gödel was exceptional.

His achievement was extraordinary in any sense. He was a person whose entire way of approaching the world was based on an extremely pure form of reason. He had the courage to follow his need for certainty wherever it led him and to accept the conclusions that he arrived at. From a world of certainty, he discovered the inevitability of uncertainty. He was a great man who glimpsed what I have been calling the science of wonder. His intuition was correct, his logic impeccable, yet he could not fully accept the implications of his conclusions because the loss of certainty is not just another abstract mathematical result. The loss of certainty cannot be divorced from one's existential and emotional reality.

It sets up a paradoxical situation. From inside the logical universe, Gödel had logically proved that there was more to reality than logic. Intellectually, you can get around this dilemma by being a Platonist; existentially you are trapped. Nevertheless, there is something at play in Gödel's life and work that is very deep.

Gödel's work is not some triviality that can be safely ignored because it is peripheral, because it supposedly has no affect on the day-to-day activities of mathematicians and scientists. Gödel was a pioneer, an explorer who ventured into some very dangerous terrain. How are we to respond to the legacy of Gödel and to the challenge that is implicit in his work?

Gödel's legacy can be approached from two points of view—either negatively or positively. The negative interpretation of Gödel's work is that any world of certainty is inevitably incomplete. Yet his work does not imply that there is no truth. Viewed positively, it says that certainty is not to be confused with truth; that the way to the truth is not exclusively through logic and deductive systems. This is precisely the import of the related work of Gregory Chaitin with which I began the chapter.

Undecidable Problems

Gödel provided mathematics with the idea that certain questions were *undecidable*. This means that a particular mathematical statement can neither be proved nor refuted within some specific axiomatic system. A specific example of such a question is the continuum hypothesis, which is a statement about the non-existence of certain kinds of infinite sets. Cantor had shown that there were two different orders of infinity involving numbers, namely "countable infinity," which is the infinity associated with the counting numbers {1,2,3,}, and the larger "uncountable infinity" associated with the collection of all the infinite decimals. The continuum hypothesis relates to the question about whether there exists an intermediate order of infinity. Gödel and the mathematician Paul Cohen proved that this question could not be decided on the basis of the usual set of assumptions or axioms that are commonly thought to support all of mathematics. But the continuum hypothesis is not the only mathematical conjecture to fall into this kind of no-man's land. It had a precursor in the parallel postulate that I discussed earlier that differentiates

Euclidean from non-Euclidean geometries. Many explicit mathematical questions can be shown to be undecidable. Gödel had shown something much more general: that *any* sufficiently complex axiomatic system, based on a manageable set of assumptions, must contain questions that are undecidable (within the system).

Mathematicians also use the word undecidable in another way. This second use of the word refers to a problem for which it is impossible to construct a single algorithm or set of formal rules that will invariably lead to a correct yes-or-no answer. A problem that is undecidable in this latter sense cannot, in general, be solved by computers. The existence and relative ubiquity of such problems says something profound about the role of computing machines and their relationship to human beings. The most famous example of such an undecidable problem is the *halting problem*. Essentially, you are given a program and some number as input. The computer takes the input and makes a sequence of computations. When it arrives at the answer, it halts. However, it may never get the answer and so will run forever. The question is to determine which one of these two possibilities happens. In 1936, the brilliant mathematician Alan Turing proved that there is no general program or algorithm running on an idealized computer, called a *Turing machine*, which solves the halting problem for all possible inputs. In other words, the halting problem is undecidable.

Today, we have a lengthy list of problems that are undecidable in this latter sense. The most famous was formulated by David Hilbert. It was the tenth in a series of problems he posed in 1900 as a challenge to the mathematicians of the next century. (My doctoral supervisor, the Fields Medal–winning mathematician, Steven Smale,[14] proposed a modern analogue of Hilbert's list in the year 2000.) The problem was to create an algorithm that would find all integer solutions (if any) to a Diophantine equation, which is a polynomial equation, like $x^3 + 3xy^2 + 2x^2y + y^3 = 0$. In 1970, the Russian mathematician Yuri Matiyasevich showed that no such algorithm is possible.

In this connection, we should also return to the work of Gregory Chaitin, who found undecidable statements in a field called algorithmic information theory and proved another incompleteness theorem in that context. Chaitin is noted for thinking extensively about the implications of his results for mathematics, philosophy, and science in general. In a way, my own thinking has deep connections with the path he blazed.

Actually, there is a deep relation between the two senses in which we used the word *undecidable*. Chaitin's work is related to Gödel's in many ways—they each produced incompleteness theorems based on a famous paradox—and also to the results of Turing. The conclusion on a metamathematical level is that incompleteness and both varieties of undecidability are different ways of talking about the same underlying phenomenon. Chaitin and those who have commented on the larger significance of his work, are very aware that there is something going on that is highly significant. Incompleteness and the undecidable are different ways in which the blind spot finds its way into science. They indicate that there are limits to what can be achieved through a certain kind of thinking. Such limits are today firmly established as part of the intellectual landscape. However, their implications are still far from having been assimilated into the general culture. In general, our culture remains in a "classical" state, yet the frontiers of science have moved on. Something fundamental has changed.

The work of Cantor and Gödel and the thinkers who followed them is not the end of the story, but merely the beginning of a line of thinking that continues to this day. The main theme consists in the rediscovery of a basic feature of the relationship between the mind and the natural world that had been unacknowledged during the epoch when Western culture was in denial regarding the inevitability of uncertainty. Once the genie is out of the bottle, it tends to turn up all over the place. This is the way it often is with fundamental scientific discoveries—when the moment is propitious, you may find the same breakthrough being made simultaneously by more than one researcher.

Wittgenstein's Silence

Ludwig Wittgenstein was introduced in the previous chapter. Like Gödel, Wittgenstein had a great deal to say about the nature of logic and mathematics, but more particularly the limits to reason. My remarks will be based on the recent book on Gödel by the philosopher and novelist Rebecca Goldstein,[15] especially her comments on Wittgenstein's early masterpiece the *Tractatus Logico-Philosophicus*. In the preface of this work, he says, "The whole sense of the book might be summed up in the following words: what can be said at all can be said clearly, and what we cannot talk about we must pass over in silence." This last phrase is also the final proposition of the book, its culmination and the context within which his work (and mine) must be judged.

The meaning of this statement is central to the content of this chapter but, strangely, that meaning has been so twisted it has come to mean the opposite of what was originally intended. It has been taken to mean that "the misuse of the conditions of language not only (tautologically) leads merely to nonsense, but also beyond the bounds of the sayable there is nothing at all, whereas for Wittgenstein there really was 'that whereof we cannot speak.'"

Again we can see that Wittgenstein's statement can be seen in both the positive and negative manner with which we have now become accustomed.

What, then, is the nature of Wittgenstein's silence? Is it blankness, an impenetrable void? Is it a potentiality, an absence of some sort? Wittgenstein is responding to the paradoxical position he found himself in—the necessity of articulating what cannot be articulated, of speaking about the existence of something that cannot be captured in words. This is the vision of a primordial Truth that exists but cannot be grasped or understood. Sometimes people who speak like this are called mystics almost by definition, and if so, Wittgenstein was a mystic.

What Wittgenstein is pointing to is no different from the realm of mystery that I referred to in the previous chapter as the "ungraspable" or the "blind spot." What makes all the difference is whether you take Wittgenstein's statement in a positive or negative way. Wittgenstein meant it positively, that there was an aspect of reality we cannot speak about in the sense that anything we say is not it. Accepting this means putting a huge hole in the middle of traditional metaphysics. Rightly understood, it precipitates a crisis, and passing through this crisis leads us to a new world.

UNCERTAINTY IN SCIENCE

Mathematics is not the only scientific domain within which uncertainty has revealed itself. I mentioned earlier that quantum mechanics is a kind of paradigm for a new, post-classical, way of looking at science. It even has its own "uncertainty principle," which states, for example, that complete and certain knowledge of the position and the momentum of a subatomic particle cannot be obtained simultaneously. Thus, certain knowledge about the total state of a physical system is impossible to obtain in principle.

Quantum mechanics also has the notion of "complementarity." Is an electron a localized particle or is it a probability wave? It is both but cannot be both—thus, there is an element of uncertainty that is implicit in the description of subatomic particles. Quantum physics does not have the same level of certainty as classical physics. The classical ideal of complete, objective, and absolutely certain knowledge has been lost. The ability of reason to "capture" the world has been put in doubt.

Much earlier, the physicist Ludwig Boltzmann (1844–1906) developed the notion of entropy in thermodynamics. Entropy is a measure of uncertainty or disorder within a system. The second law of thermodynamics states that the entropy of a system must always increase. From this point of view, not only is uncertainty inevitable, but systems tend to get more and more disordered. Certainty can be thought of as a form of extreme order. Entropy is a measure of disorder.

In earlier chapters,[16] I discussed randomness and mentioned that it was one of the most important concepts in both the theory and methodology of modern science, from statistical techniques to the theory of chaotic systems. Randomness means a lack of order, and without some kind of order it is difficult to see how there can be certainty. Many of the great theories of modern science testify to the fact that uncertainty is an irreducible feature of the world.

A word needs to be said specifically about the biological sciences. The dominant perspective here is deterministic, yet occasionally other voices can be heard. In chapter 4, I shall discuss the biologist Carl Woese's view that "a living creature is a dynamic pattern of organization...." A deterministic point of view has little room for uncertainty, nor for a view of life as "patterns of organization [that] are constantly forming and reforming themselves." A classical, deterministic science is a science of stasis. It misses the essence of life, namely dynamic change. It would seem that it is precisely the life sciences in which the need for a new perspective is most compelling. They must stop looking for a paradigm in an outdated view of the physical sciences.

CONCLUSION

This chapter concludes with a recapitulation of two essential points. The first is that a whole series of scientific discoveries in the last hundred years are all pointing in the same direction—the direction of limitations to a particular way of interacting with the world. I will use many different words to draw your attention to this situation. These words include uncertainty, incompleteness, and ambiguity, the ungraspable, the blind spot, or the limits to reason. These are all words or expressions with a negative connotation; they are written from the viewpoint of certainty and reason, from the classical point of view.

However, it would be wrong to leave matters at the stage of the classical world versus those who claim that the classical world is deficient and deluded. This brings us back to Chaitin's remarks about the significance of Ω with which the chapter began. Every crisis has within it the seeds of rebirth, and this particular crisis carries the potential for a revolutionary new way to imagine the world, the role of human beings, and the nature of the mind. The loss of certainty can be seen as the end of the world, or—as I have repeatedly said—it can be seen as an opening to a new way of thinking about science.

One of the factors that characterize this moment in history is our civilization's current confrontation with problems of such complexity and urgency that we find ourselves paralyzed. These include economic problems, vast population growth, scarcity of food and clean water, depletion of energy sources coupled with the insatiable desire of people in developing countries like China and India to share in the promised land of middle class life and the consumer economy. The problems are so vast, and the proposed solutions so stopgap, that even thinking about the problem becomes painful and seemingly futile. The problems are so basic that it is difficult to conceptualize them in any practical terms; they are, to all extent and purposes, ungraspable. Given this, it is understandable that we sometimes choose to look away.

The present world crisis, for that is what it is, will not be resolved by "business as usual," by applying techniques and strategies that have brought on the situation we are in. We must find a way to approach our cultural blind spots, to face uncertainty with courage, to confront our own powerlessness in the face of what I have been calling the ungrasp-able without becoming paralyzed. Our situation contains elements of the paradoxical—it is the successes of the scientific method that have brought us face to face with its limitations, just as it is the complex-ity of the global financial community that has brought on the present economic meltdown. Our best response at the present time seems to be more of the same way of analyzing problems that got us here in the first place, whether it be the latest technological fix or more standard financial and economic models. At times, it seems hopeless, like there is nothing we can do. But we must do something! This is our collective encounter with the blind spot. Facing the situation authentically will evoke the cre-ative intelligence that we desperately need at the present time.

3

⌢∿

Certainty or Wonder?

What are the implications for science and society of the existence of a blind spot that cannot be captured by systematic thought? We saw in the last chapter that there can be two reactions, which I characterized as negative and positive. In the former, the emphasis was on the fact that reality is forever uncertain, that all systems of thought have their limitations. The latter involved the realization that uncertainty and incompleteness are the price we pay for creativity—in fact, for being alive.

In this chapter, we shall see that these two points of view lead to two different orientations toward the entire scientific enterprise. One I shall call the "science of certainty," and the other the "science of wonder." They differ in many ways, but one crucial difference involves their attitude toward the uncertainty and incompleteness that the previous chapters discussed. Now, to divide up science in this way is a little simplistic and unfair since I would say that all of science, deep science, is the science of wonder. It emanates from a sense of the wonder and mystery of the natural world that is the subject of the first part of this chapter. The demand for absolute certainty is more characteristic of a mythology of science, a view that one finds more often in technology than in pure science. It is a distortion of pure science, a distortion that, as we shall see, has the potential to create a lot of damage.

In other words, there remain different ways in which science can be viewed with vastly differing consequences. It is essential that we learn to differentiate between them.

SCIENCE AS WONDER

To the natural philosopher there is no natural object
unimportant or trifling ... a soap bubble ... an apple ... a
pebble ... he walks in the midst of wonders.
—John Herschel[1]

The most profound motivation for doing science is the sense of wonder, even awe, that arises naturally when confronted with the intricacies of the natural world. From the workings of the cosmos down to the inner structure of a grain of sand, the world is filled with an extraordinary beauty that evokes a sense of wonder in those most receptive to it. John Herschel, the nineteenth-century mathematician, scientist, and photographer, captures the sense of what has been called "romantic science" in the preceding quotation. As the historian Richard Holmes points out in his recent book, *The Age of Wonder: How the Romantic Generation Discovered the Beauty and Terror of Science*,[2] "Romanticism as a cultural force is generally regarded as intensely hostile to science, its ideal of subjectivity eternally opposed to that of scientific objectivity. But I do not believe this was always the case, or that the terms are mutually exclusive. The notion of *wonder* seems to be something that once united them, and can still do so." Wonder has never been absent from science, but it disappears when we turn science into something mechanical, or use it as a way of achieving total control over nature or society.

It is interesting that Richard Dawkins, the well-known evolutionary biologist and writer, was in agreement with Herschel when he wrote, "The spirit of wonder ... is the very same spirit which moves great scientists; a spirit which, if fed back to poets in scientific guise, might inspire still greater poetry."[3] Even though Dawkins is known as a very hard-line scientific rationalist,[4] he is also a secular humanist and, as such, also has a prominent place for wonder in his view of the world.

The very same point is made in one of Einstein's most enigmatic and delightful statements, "The most beautiful experience we can have is the mysterious. It is the fundamental emotion that stands at the cradle of true art and science."[5] What does Einstein mean by the mysterious? Science ostensibly appears to be about qualities like certainty, precision, objectivity, clarity, and predictability. These qualities have nothing mysterious about them. On the contrary, they seem to be as far from the mysterious as it is possible to go.

A discussion of science should properly be based on the centrality of wonder, mystery, and creativity. The certain, the precise, and the predictable will then be seen as secondary characteristics that should be viewed within a larger context. It is this expanded context that evokes the sense of mystery and wonder. This is where science comes from and where we should look if we want to understand science in a deeper way. Mystery and wonder are inseparable from that grand adventure of human creativity that is the correct perspective from which to think about science.

The "science of wonder" is based on Einstein's sense of the mysterious. It contrasts with "the science of certainty" that will be discussed later. The misconception that the "science of certainty" is identical to science has contributed to many of society's current problems. That statement should be nuanced because it seems to place these two points of view on an equal footing. Actually, the "science of certainty" is not the polar opposite of the "science of wonder." It is *derived* from the science of wonder. That is why Einstein speaks of the mysterious as being at the *cradle of true ... science*. The science of wonder has room for both certainty and uncertainty. Thus, it is appropriate for us to begin with the larger context; the science of wonder.

Now, you don't have to be a scientist to appreciate the science of wonder. The capacity for mystery and wonder is a universal human potential. You could even say it is the deepest manifestation of what it means to be human. Take, for example, the experience of reading a book about science like *The Elegant Universe*, Brian Greene's wonderful book about string theory[6] or even watching the *Nova* program based on it. Both book and program were fascinating and moving. What this theory is attempting to accomplish—nothing less than to establish a mathematical framework within which to understand the workings of the whole universe—is something noble and audacious!

What exactly is it that is so moving about such scientific theories? Forget for a moment the technical details, the content of the particular theory, and the applications even if they have the capacity to change the world. None of these explains what is so profoundly moving. There is something about certain presentations of science that are completely captivating, as though this theory was letting you in on a secret of cosmic significance—some organizational principle of the universe. You may only understand it in a very incomplete way, yet it is very exciting to feel that you are privy to a blueprint of reality; that you have been given a deeper insight into the way things are. You now look at a part of the natural world in a way that was formerly unexpected. New connections have been revealed and therefore a light has suddenly illuminated something that was formerly obscure. Creativity is connected to, even based upon, illumination. Where does illumination come from? What are its sources? One thing is certain. Whatever we mean by illumination, it is intimately connected to the mind, and therefore to intelligence.

Now such an insight—even if it is second-hand—combined with the sense of wonder, is actually a creative experience in itself. What happens at such moments is that the individual has a "Eureka" moment. It is not different in kind, although certainly in depth, from the creativity of the great scientists. A great expositor can communicate to the reader, listener, or watcher something of what it means to be creative in science. If you are sensitive to such things, the feeling that arises is very powerful. Normally, we only pay attention to the theory that evokes such a response, but in this book the response itself will be seen to be of immense value. However, the creative experience I am interested in is not reserved for people of genius. Creativity is the common heritage of everyone. How does this universal tendency reveal itself in science? What does science look like if we consider it to be intrinsically connected to human creativity and not as absolute truth residing in some objective domain that is separate from human intelligence? In order to take a fresh look at the nature of science, I shall have to temporarily separate the creative aspect from the content of the particular mathematical or scientific theory that induced the experience.

When I say that I am interested in talking about creativity, I am not implying that this is a psychological study. I do not wish to divide science into the content of science, on the one hand, and the individual who creates the content, on the other. In my view, theory and process cannot be definitively separated from the scientist who creates them, and even from the student or layperson who learns about them.[7] All of this, the content and process, is part of the vast field of creative endeavor that I shall be referring to when I talk about science. I am not so much interested in the psychological phenomenon of creativity as I am in investigating what creativity in science tells us about the nature of the subject itself and what looking at things in this way tells us about how human beings operate as they try to glean correct inferences about the natural world and themselves.

Now let's go back to the feeling that many of us experience in the face of an inspiring exposition of deep scientific ideas—the feeling we have been let in on some secret of cosmic significance. Such a feeling was once more often associated with religion than with science, and this may be one reason why science has been called the religion of our time. Religion once provided what science provides today—a vast mythological structure that contains an explanation of the cosmos and the particular role of human beings in that cosmos. Religion, through its rituals, its art, and its stories, provided people with meaning but also with a sense of awe and mystery. Science also provides access to these feelings. Science provides the sense that the universe is a coherent, connected, and meaningful entity—a sense of the unity of things. This sense of coherence is powerful and important even if most people believe it is merely a subjective sensation and therefore a derivative feeling of secondary importance. Coherence is important precisely because it has both subjective and objective dimensions; because it is related to and accompanies acts of creativity. The sense of wonder often accompanies the sudden appearance of a sense of coherence.

Wonder, since it is derived from meaning, coherence, and creativity is a good place to start building up a new picture of what science is all about. This new picture will make a start toward closing the gap between the subjective and objective that needs to be closed for the sake of our well-being as individuals, as well as for the sake of the health of the global society as it faces the enormous challenges of our time. The alternative is a continued alienation from the natural world and even from groups of human beings that do not belong to our "tribe," be they religious, ethnic, or economic. Coherence results from acts of creativity, but coherence is itself a sense that unifies the natural world with the human mind as it grapples with, and attempts to make sense of, that world. Thus, coherence, and therefore creativity, is something that transcends classical objectivity and subjectivity.

THE SENSE OF WONDER AS
"COSMIC RELIGIOUS FEELING"

In a well-known article that first appeared in the magazine section of *The New York Times*, Albert Einstein introduced the expression, "cosmic religious feeling" when referring to science and scientists. The article was titled, "Religion and Science" and is really an expansion of his comments on the mysterious:

> ... there is a third stage of religious experience ... even though it is rarely found in a pure form: I shall call it cosmic religious feeling. It is very difficult to elucidate this feeling to anyone who is entirely without it, especially as there is no anthropomorphic conception of God corresponding to it.
>
> The individual feels the futility of human desires and aims and the sublimity and marvelous order which reveal themselves both in nature and in the world of thought. Individual existence impresses him as a sort of prison and he wants to experience the universe as a single significant whole....
>
> How can cosmic religious feeling be communicated from one person to another, if it can give rise to no definite notion of a God and no theology? In my view, it is the most important function of art and science to awaken this feeling and keep it alive in those who are receptive to it.

How is one to bring to people's attention, Einstein asked, the crucial importance of a feeling that is in essence religious, yet not employ an "anthropomorphic conception of God"? That task is no easier today when society seems evenly split between those who do and those who do not hold to such an anthropomorphic conception—when religious people view science with suspicion and many scientists see religion as mere superstition. The problem here is that the argument is mistakenly about the *content* of science versus the *content* of religion. Einstein's "cosmic religious feeling" is meant to bridge that particular gap—to show that the highest form of religious feeling is to be found in science as well as religion. (Of course, just calling a certain point of view religious or scientific does not ensure that "cosmic religious feeling" is to be found there.)

"Cosmic religious feeling" is the unifying context from which both science and religion emerge and is intrinsically connected to creativity.

Einstein, in this excerpt, is revealing the personal origins of his scientific activities—why he is a scientist. Not only does cosmic religious feeling inspire him to do science but also *doing* science evokes this feeling. Now, by doing science I mean especially to emphasize those creative moments when what was formerly an intractable problem suddenly comes into focus and a new way of understanding is born. Thus, cosmic religious feeling is both an inspiration for, and a result of, creative activity in science. It accompanies the great scientist at every step of his or her journey into the unknown. I maintain that Einstein's words also apply to mathematics, that the same "cosmic religious feeling" also motivates mathematicians. Therefore, in this chapter when I use the word science I will also be talking about mathematics. From the point of view of the subjective experience of the scientist, there is no difference. It is merely the domain that evokes this feeling that is different—for the biologist, it is the biological world; for the physicist, the physical world; and for the mathematician, it is the mathematical world.

Without taking into consideration Einstein's "cosmic religious feeling," by merely focusing on the content of his work without including the ground from which the work emerged, you run the risk of missing an aspect of Einstein's being that is essential to an understanding of Einstein as scientist and human being. Thus, it is important, but not only for understanding what makes a great scientist tick. "Cosmic religious feeling" is not reserved exclusively for the great scientists. Recall that he says that, "it is the most important function of … science to awaken this feeling and keep it alive in those who are receptive to it." Those of us who love science and are captivated by it; those of us who find that science awakens in us a sense of awe at the mysterious yet comprehensible workings of the universe; we are the ones who are receptive to Einstein's "cosmic religious feeling." We thereby share this basic human potential with the great scientific geniuses.

In Einstein's view, science and religion both arise out of this common ground—the same cosmic religious feeling. This is what we all share with Einstein and, potentially, with all of humanity. In the seventy odd years that have passed since Einstein wrote those words, the split between religion and science has deepened and many people who are interested in science might even object to mentioning religion and science in the same paragraph. For such people, it would perhaps be helpful to substitute the word "spiritual" for the word "religious" so as to emphasize that Einstein was not referring to some organized body of religious theory but to a feeling, a very common human feeling, that we could also call a sense of beauty. Great science emerges out of this sense of awe and, in its turn, evokes it in others.

The order and precision of science and especially of mathematics stand against the inchoate world of everyday life, and it is this comparison that makes science so attractive. Nevertheless, the order of science emerges out of something that is even more basic, some ineffable realm that we can only know by inference. The emergence of that which we do not know into the light of reason is accompanied by a sense of awe, or of cosmic religious feeling. Today, basic science is a primary area within which people may access the sense of awe that accompanies the mystery that marks the emergence of the unknown into human consciousness. As Einstein says, in studying science we are simultaneously studying religion in its deepest sense.

Normally when one talks about Einstein, one discusses the content of his work—the theory of relativity, for example. Wonder, this "cosmic feeling," is never even alluded to. And yet I maintain that this feeling needs to be talked about and understood, for it too is part of science, and without its inclusion, elements of the scientific enterprise remain incomprehensible.

When we lose ourselves in the contemplation of nature or of a problem in science and suddenly discern a relationship in a situation that previously made no coherent sense, at that immediate moment we experience an exquisite sensation for which there is really no adequate verbal description. Why is this feeling so seductive? Why do people, once having had this experience, keep attempting to re-create it? I maintain that Einstein is pointing to the most valuable experience that a human being can have (he clearly regards it as such himself)—the sense that in these moments of clarity we see things as they really are. For one moment, a certain illusory veil drops away and we find ourselves at home and at ease in the universe. We have the feeling that we have never been so alive—that things have never been so "right" as at this moment. In this sense, creativity is self-validating. Having experienced this moment we are now consciously aware of the existence of such a state, and something in us will spare no effort to return there. I am not saying that the scientist is not fascinated by the objective research problem that she finds herself working on but that, in addition, there is, consciously or unconsciously, a desire to return to this subjective state that accompanies acts of creativity large and small. Normally in any description of science, we proceed by assuming that the objective theory can stand independently of any "subjective" component—this is considered to be the very defining characteristic of science. Yet I shall ground my description of science in human creativity. This will not denigrate science—in fact, it will allow us to clearly discern the genius of the scientific enterprise. Viewing science as a field of creative endeavor will allow us to think of science in a radically new light, which will then reflect back on our understanding of the human condition itself. Normally, we think of this light as only traveling in one direction—that is, science is seen as providing an explanation of the deeper properties of the natural world and therefore of human beings. I shall turn this around and think about what the nature of scientific activity and of scientific culture tells us about what it means to be human.

"Cosmic religious feeling" is intimately connected to creativity. When a person has one of those creative moments when things "click," there is, for a brief period of time, a sense of wholeness and well-being that arises from breaking through a primordial tension that, as we shall see, is intrinsic to the human condition. The scientific problem that is resolved stands in for the larger existential problem that the scientist is living. If the existential problem is not resolvable in any conventional sense, the scientific problem can still sometimes be solved and, in so doing, the larger burden is, if not resolved, at least made lighter. Thus, I am not interested in any sense of victory or incipient fame that may arise in the author of a scientific breakthrough. I am interested in the sense of "knowing" or of clarity that arises in the scientist. Interestingly, this reaction is instantaneous and precedes any thought of personal advantage. The latter all-too-human reactions are really a contamination of the more immediate opening that is not so much a reaction to an act of creativity as it is a part of the creative experience itself, which therefore arises with it.

I want to talk about the incredible creativity of science in a way that highlights that creativity, not as a means to an end but as an end itself. There would be no science without human creativity and that creativity should be part and parcel of what we are referring to when be we talk about science. However, we must be careful here not to slip into the belief that science is completely created by the scientist. In the paragraphs that followed the excerpt I quoted earlier, Einstein referred to the great scientists' "deep conviction of the rationality of the universe and [the] yearning to understand [even] a feeble reflection of the mind revealed in the world." Normally, people do not take Einstein seriously when he talks about things like "the mind revealed in the world"; they ascribe it to a cute, metaphoric turn of phrase. Yet perhaps Einstein should be taken more seriously when he senses that there is an intelligence that is revealed in the workings of the universe. It is not that human intelligence creates an order in what is essentially a chaotic domain but that some intelligence or "mind" is an intrinsic aspect of reality. What a radical challenge to our usual way of seeing things! The scientist reveals that universal mind to us, but that mind is there independent, in some sense, of the individual. Einstein talked about "God," while his good friend, the great logician Kurt Gödel, talked of the reality of the world of mathematical objects. They did this because for them this intelligence appeared to be independent of the individual personality, independent of anything that is personal. What is exciting and so mysterious is that this "mind … in the world" is reflected in the human mind. This larger external intelligence is accessible to human intelligence. This is the true mystery of science. Creativity is not the imposition of arbitrary patterns on the world. It consists of making explicit an order that is implicit, as the distinguished theoretical physicist and scientific thinker, David Bohm, would say. It consists of revealing an order that resides neither in a realm that is completely objective nor in one that is subjective. It reveals one complex reality in which the so-called objective and subjective are merely two different points of view.

Having now established the sense of wonder at the center of scientific creativity, I shall go on to compare and contrast this feeling with the sense of certainty and precision that characterizes scientific thought. Which of the two, wonder or certainty, is more basic? Is it possible to have certainty without wonder? And if so, what are the consequences?

THE NEED FOR CERTAINTY

He said we could face the worst if we simply renounced
our yearning for certainty. But who among us is
capable of that renunciation?
—Michael Ignatieff[8]

Reality, as you can tell from the daily news or from the events in your own life, is uncertain. Science, on the other hand, is intimately connected with the human "yearning for certainty" that Ignatieff writes about. If we wish to understand the scientific enterprise in all of its strengths and weaknesses, we could do no better than to begin with our own yearning for certainty. The philosopher, logician, writer, and Nobel Prize winner, Bertrand Russell said, "I wanted certainty in the kind of way in which people want religious faith. I thought that certainty is more likely to be found in mathematics than elsewhere."[9] Russell's statement applies to science in general and not only to mathematics. Science is *the* place that most people look to satisfy the human basic need for certainty. Scientists may well feel the need for certainty more acutely than other people, but the "yearning for certainty" is a basic part of the human condition. We all share it.

So what is the alternative to certainty? For many people, it is the chaos described by Shakespeare in *Macbeth*, "it [life] is a tale told by an idiot, full of fury, signifying nothing." The "tale told by an idiot" is a chaotic world devoid of meaning and coherence. This is the stuff of nightmares, and science seems to offer the possibility of waking up from that nightmare into a world of clarity and objective certainty. Certainty is surely important both socially and psychologically, but certainty overdone leads to rigidity and resistance to change. There is a trade-off here. In our desire to satisfy our need for certainty, we may end up sacrificing deeper human requirements, such as freedom and creativity.

WHAT IS CERTAINTY?

One dictionary meaning for certainty is the absence of doubt. René Descartes was the philosopher and mathematician who famously claimed to doubt everything in his attempt to secure a solid foundation for knowledge and thought. He found that he could doubt everything *except* the doubt itself. Doubt thus became, paradoxically, the foundation for his system of certain knowledge.

Science promises certainty. This is what drew the novelist and physicist Alan Lightman to science. He wrote, "Mathematics contrasted strongly with the ambiguities and contradictions in people. The world of people had no certainty or logic." Lightman and Bertrand Russell both had an enormous thirst for certainty and I believe this is one of the most basic things that drives people into scientific careers. When philosophers look for certainty, the models they seek out are the sciences. But it is not only philosophers, everyone who uses science and mathematics, including engineers, businessmen, actuaries, financial analysts, and civil servants make use of the certainty that is associated with math and science to support them in the decisions they must make every day.

When I look back for the reasons why I was initially attracted to mathematics, this dimension of certainty was prominent. The promise of "knowing beyond all doubt" was so attractive. What a great feeling it is—to know and know that you know! You want to preserve these jewels of certainty because they seem to have so much value. In fact, the qualities of certainty and value seem to go together. In ordinary life, on the other hand, every argument seems to evoke a possible counterargument. Oftentimes, it seems impossible to make a definitive judgment without using other, equally questionable, assumptions. I am reminded of a famous story of a rabbi who was presiding over a rabbinical court, before which two people were pleading their case. The first person made such a convincing argument that the rabbi could not prevent himself from exclaiming, "Yes, you are right!" Unfortunately, the second plaintiff was equally convincing, and so the rabbi was forced to say, "You know, you are right!" At this point, the clerk of the court intervened to say, "But rabbi, they can't both be right." And the rabbi answered, "Yes, yes. You're right." Life seems to be like this, filled with conflicting positions that seem to have an equal claim to validity—filled with ambiguity. To sort out the right from the wrong, to make decisions in the midst of confusion and have confidence in the correctness of the actions that follow from those decisions, is difficult and confusing. How many of us long for a world in which the ideal of certainty is attainable!

This is the world that science seems to offer us—a world of objective certainty. Science provides society with a vision of an existence in which doubt has been vanquished and certainty reigns supreme. In this vision, not only is this world of certainty the factual state of affairs—the way it is—but the scientific method is the unique way in which that objective can be attained. Through its marriage of the empirical and the rational, science appears to have carved out a domain of certainty in our interactions with the natural world. Moreover, it claims that this domain of certainty can be expanded to include just about everything.[10]

Now you might say that, on the contrary, what is important in science is curiosity. It is indeed a great motivator and every mathematician or scientist is curious, but curiosity needs to be resolved. Questions are not enough. We must have answers! We must have certainty! In mathematics, certainty is guaranteed by the existence of a proof. Thus, mathematical theorems epitomize certainty. A mathematical theorem appears to be at the other extreme from a merely subjective phenomenon. This is the reason why so many people love mathematics—it doesn't change. It seems to be true once and for all, for every person at all times. We say, for example, "it is as certain as $2 + 2 = 4$." The essence of certainty lies in its claim to objectivity.

The traditional method for attaining certainty involves the use of logical reasoning. When we go through a proof of the Pythagorean theorem, we feel that the matter is settled; the result is true and will always be true. Unfortunately, the promise of absolute certainty can play us false, as Reuben Hersh and Vera John-Steiner point out in *Loving and Hating Mathematics*.

> It is really claimed by some philosophers that the propositional and predicate calculi—modern formal logic—are infallible (e.g., John Worrall and Elie Zahar in editing Lakatos's *Proofs and Refutations*). "From true premises, true conclusions follow, infallibly." How dangerous this dogma can be. Logic can never be anything but a tool, or an action, or a procedure, carried out by a *human being*. (Or by a machine created and programmed by a human being.) Logic, such an essential tool of science and philosophy, sometimes becomes a sort of false god, outranking the most fundamental human impulses, such as "Thou shalt not kill."[11]

In our search for an absolute certainty, logic is often divorced from its human origins and given an independent existence above and beyond the human experience. Unfortunately, logical thought can be misused in the pursuit of dubious aims. It is not that logic ensures certainty but that logic arises out of the human need for certainty. This is not to say that logic is not a useful tool, and is not invaluable and essential to both science and mathematics. But logic is not primordial and cannot ensure absolute and objective certainty. The relationship between logic, certainty, and truth is very complex. For the moment, let us just say that logic is the methodology of certainty in science and math. Through logical reasoning we hope to establish a permanent and objective state of certainty.

The Paradox of Certainty

Further thought reveals that the whole idea of certainty is a little strange, even paradoxical. Though certainty is identified with objectivity, what often passes unobserved is that certainty is a *feeling,* sometimes referred to by using a negative criterion, the absence of doubt. However, certainty is not an absence. It is a feeling that has elements of assurance and conviction, of firmness and strength. It is a subjective state that arises in the mind. Sometimes the subjective aspects of certainty are distinguished from the objective through the use of the word "certitude" rather than certainty.

However we must we use the word "subjective" very carefully. When I talk of certainty as subjective, I do not mean it is merely a matter of opinion. Certainty can arise in many different scientific situations, but it is the *same certainty* in all cases. It is merely the context within which the certainty arises that changes.

The essence of certainty is that it contains both objective and subjective aspects. How can a subjective state of mind arise from one's interaction with a domain that is totally objective? This is a great question, which I shall return to later. For now, let's try to appreciate the subtlety of the phenomenon of certainty. To be *certain* is to have the subjective feeling that something is objectively true. Isn't that peculiar? Isn't it ambiguous? Strangely then, though certainty promises to do away with ambiguity, certainty itself is ambiguous. This kind of irreducible ambiguity will be discussed in subsequent chapters.

The search for certainty and the stability it implies is a fundamental human drive. Without the certainty that our physical environment is secure, and that our important relationships are solid, life is very difficult. One reads stories of people who have lived in situations of war where there is no certainty that the essentials of life—food and shelter—will be available from one day to the next. In addition to their physical privations, such people endure incredible stress. So even in ordinary life certainty brings with it a sense of security and well-being, a feeling of being in control. Uncertainty is painful; the psychoanalyst Robert Winer even talks about "the human intolerance for uncertainty"[12] as a very basic aspect of life. There is a dilemma here built right into the heart of most people's lives. It consists of this intolerance for uncertainty coupled with the recognition of its inevitability and ubiquity. Science is one stage on which the drama generated by this dilemma is played out.

The research environment is not only a domain of certainty. A common research observation is that for every answer laboriously attained, a thousand new questions inevitably arise. However, the layperson believes that the aim of science is to build up a domain of irrefutable certainty and that this domain is the real science. For this reason, the contents of science are believed to exist independently of the activity from which this content is produced. It is our need for certainty, permanence, and objectivity that causes us to divide up science in this (artificial) way. Think of the expression the "laws of nature." Why do we use the word "laws" and not "patterns" or "regularities"? Surely to say that something is a law is to maintain that it is absolute, immutable, certain. Yet it is conceivable that even these "laws" change; that they too are governed by the principles of evolution. Is the role of science to discover the "laws of nature" or the "truths of nature"?

When the mathematician Paul Halmos was asked, "What is mathematics?" he answered, "It is security. Certainty. Truth. Beauty. Insight...."[13] The first two characteristics—security and certainty—represent one approach to science. "Truth, beauty, insight" represent another. For most people, science is science; it is monolithic. For me, there exist conflicting tendencies within science and I feel that it is very important to distinguish between them. Doing so will help us understand how science can be both a blessing and a curse; how it can contribute to the solution of many social problems, on the one hand, and, on the other, how a misunderstanding of the nature of science may create new problems and exacerbate those we already have.

Again, there are these two ways of looking at science—wonder versus certainty. The science of wonder springs from the sense of awe, from Einstein's cosmic religious feeling that I discussed earlier. It arises naturally from some interactions with the natural world, but also from our interactions with the cultural world, from great works of art and music, from deep theories of mathematics and science. It is characterized by a limitless openness and creativity. It emphasizes process over content, and has elements of objectivity but also of subjectivity. It is never over, never totally completed, nor absolutely certain. "Wonder" is a way of approaching science that differentiates it from the approach that I've been calling the science of certainty.

Everything Is Certain

Sometimes the tendency toward capturing certainty in science is pushed to an extreme, as in the notion of what is called the "theory of everything." Theories of everything have been proposed from time to time in the history of science. During certain moments of history, some scientists have felt they understood the theoretical basis governing all phenomena, and all that remained was to fill in the details. Today, the term "theory of everything" usually refers to some conjectured theory in theoretical physics that explains and unifies all that is known about physical phenomena. The prime candidate in recent years has been "string theory" or its sequel "M-theory."[14]

The belief in such a theory and the search for it has deep roots in our cultural history. It is a dream or a foundation myth for our culture going all the way back to the ancient Greeks. Euclid developed a system of thought that promised to make certain all the truths of geometry, an approach that was so successful later thinkers tried to extend it to many other disciplines. Laplace and his contemporaries believed that if you knew the location and velocities of all the particles in the universe, then the future and the past would be completely determined by the laws of mechanics. The same dream is at work here—every aspect of the universe can be captured by the rational mind—everything potentially resides within an ever-expanding domain of certainty.

Yet science cannot be totally identified with the felt need for absolute and total certainty, nor is religion identical to a total belief system.

Leonard Cohen said in one of his songs, "There is a crack in everything. That's how the light gets in." A theory of everything has no room for light because there are no cracks. One way of denying the existence of cracks is through the fundamental assumption that *everything is knowable*. This is the feeling that reality consists of a collection of facts and principles that can be understood completely and definitively. Sometimes people imagine that the collection of all possible data is infinite, but that the set of explanatory principles that generate this data is finite and can be made explicit. Here the words finite and infinite are used in two distinct senses—ones you could call *breadth* and *depth*. Breadth refers to the number of data points; depth to our understanding of any particular data point, an atom, say, or a concept like time, space, or number. You could summarize these two dimensions by asking whether you can know everything, on one hand, and whether you can know anything (completely), on the other.

An essential aspect of the science of certainty is the belief that there is no part of reality that is inaccessible to reason, that everything can be known—a belief in a potentially all-encompassing certainty. This is a relatively modern phenomenon and marks a decisive break with the convictions of almost every other human culture. In the fascinating book, *The Spell of the Sensuous*, the ecologist and philosopher David Abram compares scientific culture to the culture of many indigenous peoples with their emphasis on the direct experience of natural phenomena:

> In a society that accords priority to that which is predictable and places a premium on certainty, our spontaneous, preconceptual experience, when acknowledged at all, is referred to as "merely subjective." The fluid realm of direct experience has come to be seen as a secondary derivative dimension, a mere consequence of events unfolding in the "realer" world of quantifiable and measurable scientific facts.[15]

As a result of the emphasis on certainty, you often hear the claim that science has dispelled mystery from the world. However, this does not apply so much to science as it does to a particular view of science. A major theme of this book is that the view that every aspect of things is accessible to reason is not an inevitable aspect of science; that science is not incompatible with mystery. On the contrary, mystery and wonder describe the very realm out of which true science emerges. Mystery entails at the very least the recognition of uncertainty.

However the popular image of science in our culture has entrenched the idea that reality can be known completely and reduced to a collection of facts and principles, each of which is potentially accessible to reason and the scientific method. The premise that reality is finite in this way, that it can be totally brought under the sway of reason and so be made totally certain and secure, is an assumption that characterizes the culture of certainty.

What, then, is the relationship between certainty and truth? Can we distinguish between them? Certainty is permanent, but is the truth necessarily permanent? Is certainty an irreducible feature of the world, or is it a projection of a human psychological need? If uncertainty is inevitable in life and in the natural world, how is that reflected in mathematics and science? These are very important questions and the answers are subtle.

The Shock of Uncertainty

Science is attractive on a psychological and sociological level in part because it seems to promise certainty, stability, and control. What happens then when science and mathematics discover uncertainty and incompleteness as an intrinsic and unavoidable feature of the world? The initial response is usually a denial that the uncertainty really exists at all. You get this feeling in the argument between Einstein and Bohr about the nature of the world of subatomic particles. Is uncertainty inevitable and inescapable (Bohr) or is it due to a lack of understanding that will one day give way to a more complete theory (Einstein). They were asking whether uncertainty is an intrinsic part of reality or merely a feature of our current description of the world. Reality is uncertain—no amount of scientific progress will succeed in removing uncertainty from the world. Yet just saying that reality is uncertain does not do justice to the complexity of the situation. Reality is certain, too. The islands of relative certainty that science has carved out are of immense importance and the scientific method itself is one of humankind's prime hopes for the future. Certainty is not an illusion, but certainty is not absolute nor totally objective. This book is in many ways an attempt to unravel the implications of this situation.

As I discussed in the last chapter, one of the momentous events in the science of the last century has been the recognition of uncertainty as a ubiquitous feature of the scientific description of reality. But uncertainty does not arrive on the scientific scene as just another ordinary feature of science. Uncertainty arrives as a revelation and not a pleasant one. It poses a threat both on a personal and social level.

Today, many scientists would accept the existence of some irreducible uncertainty, yet as a society we are still at the early stages of coming to grips with the implications of the uncertainty gap within science—that black hole in the center of what was seemingly the most secure element of our culture. For the most part, we remain in a state of denial. But the genie of uncertainty and insecurity has been released from the bottle and there is no way to get her back in. Uncertainty is always present. There have been times in the past when the dominant culture was strong enough to ignore uncertainty, but this is not true today. We live in challenging times, and uncertainty is at the heart of the challenge.

I have already indicated that science is not monolithic by talking about two different orientations to science, which I characterized using the terms certainty and wonder. Another way to differentiate between these two aspects of science is to talk about science with or without self-reference; the latter I will call classical science as opposed to the former, a more complex kind of science that I shall discuss in chapter 6. The distinguishing feature of classical science is a kind of unambiguous objectivity and certainty. Modern science is replete with results that force us to face ambiguity and "the limits of certainty," the limits to what can, even in theory, be known.

Uncertainty is an inevitable part of the world as described by Boltzmann and Gödel, by quantum mechanics, chaos theory, and the theory of evolution. Uncertainty does not necessarily arise out of ignorance. It will not be removed by creating better theories. It is part of any description of the natural world. Uncertainty is real. It is the dream of total certainty that is an illusion.

As far as we know, everything changes, evolution is inevitable. Certainty refers to a state of affairs that is fixed. Is it conceivable that the truth can change? If so it wouldn't be an objective truth. Einstein's sense of the mysterious is a property of an authentic encounter with the truth, but the sense of the mysterious is not attached to a particular fact or theory. Nothing strips a situation of its sense of mystery so much as the feeling that it is pinned down once and for all.

Now certainty is so attractive that many people think that mathematics and science involve the search for a kind of absolute certainty, which is often what is meant when we talk about the "truth." I believe that the certainty that accompanies acts of creativity was so seductive that it was isolated from creativity as a value in its own right—for many it was *the* value. Logic, proof, and deductive thinking—instead of being seen as add-ons to acts of creativity—came to be seen as ends in themselves. This was the birth of the worldview that came to dominate our culture—the search for absolute certainty, absolute truth, a theory of everything. Certainty, considered as a basic drive of our civilization, is the attempt to replace the contingency of life with an ordered, static, and objective world that would not require further acts of creativity because everything was not only pinned down, but even algorithmic.

CONCLUSION

Two basic tendencies are at work today in science and mathematics. One I have described using the words creativity, illumination, and insight. The other involves certainty and objectivity. Now "doing" or understanding science involves both of these dimensions more or less simultaneously. There is a continuous pressure in the history of mathematics and science that involves eliminating the contingent, the tendency toward a more mechanical or technological science. Today, this might involve pinning one's hopes for the future on some kind of ultimate super-computer or even imagining that the right way to think about the mind is as a kind of computing device.

The whole discussion of certainty would gain by a consideration of recent developments in brain science. If the brain is a computer, then it is a most peculiar type of computer, one that is continually changing and developing. This contemporary view of the relationship between mind and brain goes under the name of "neuroplasticity,"[16] the brain's ability to modify itself. If even thinking and imagining have the power to modify neural circuits, then we shall have to review our set of assumptions about the nature of scientific activity and of the nature of science itself. If the brain is not fixed and unchanging, if the world itself is in a continual process of dynamic change, then perhaps we must begin to think about science in a new way, a way that is consistent with this dynamism as opposed to giving way to our psychological need for certainty and stability.

The idea that creative change can ever be eliminated is a dangerous illusion, yet it is a basic myth of our culture. The arts and the sciences are both creative activities—they stand on an equal footing in that regard. What differentiates them is precisely this mythology that human beings will one day get to a stage where things are definitively known and therefore fixed for all time—that a time will come when we can rest from our labors and the creative work of scientific research can come to an end. On the contrary, science is a humanity; it is a creative art form. Mathematics and the sciences are themselves subject to the laws of evolution that govern the biological world. Every step in the evolution of science is accompanied by bursts of creativity. Such certainty as we manage to attain refers to a given moment in time. It is not permanent and will inevitably be replaced by further and hopefully deeper insights into the nature of the natural world. Certainty is not absolute; there is no objective body of definitive knowledge. If there is anything that is unchanging, it is the act of creativity itself, and the certainty that it brings in its wake. Focusing on certainty to the exclusion of creativity is putting the cart before the horse. It is ignoring what is essential in favor of what is peripheral.

4

∽

A World in Crisis!

All I can say is, beware of geeks bearing formulas.
—Warren Buffett[1]

INTRODUCTION

What is the connection between our understanding of science and the crises that society is now facing? It is not that science is responsible for these crises but rather that a misguided view of science has been used in an attempt to create an environment that is secure and predictable in situations that are inappropriate. Human beings have a basic need for certainty. Yet since things are ultimately uncertain, we satisfy this need by creating artificial islands of certainty. We create models of reality and then insist that the models *are* reality. It is not that science, mathematics, and statistics do not provide useful information about the real world. The problem lies in making excessive claims for the validity of these methods and models and believing them to be absolutely certain. It is the claim that the current theory has finally captured the definitive truth that leads to all kinds of difficulties.

You could look at the problem as a fundamental misunderstanding of the nature of randomness. This is what Nassim Nicholas Taleb, a Professor in the Sciences of Uncertainty at the University of Massachusetts at Amherst, says in a recent book with the title, *Fooled by Randomness*.[2] Randomness is unpredictable by definition. The evolution of species, societies, and ideas contain elements of randomness that just cannot be eliminated. Defining randomness by putting it in equations and theories is a dangerous game. Such attempts are admirable and sometimes brilliant, but they are dangerous when they make us believe we have tamed randomness permanently. The improbable, what Taleb calls the "Black Swan"[3] event, is inevitably waiting for us around the next bend. Pretending that scientific progress will eliminate uncertainty is a pernicious delusion that will have the paradoxical effect of hastening the advent of further crises.

Science arises in part from the human need for certainty but it also amplifies this need by giving us the feeling that uncertainty can be definitively banished. We must remind ourselves that the need for certainty is only part of the story that is told by science. The other part involves our even more fundamental need for freedom and creativity, more accurately, a need to express our *innate* freedom and creativity.

Science can be used to provide a mythological structure that hides what is real. It is this mythological structure that does the damage by restricting the range of our possible responses to the problems that we face. In what follows I shall sketch how the scenario I have just described plays itself out in two current crises: the economy and confrontation between fundamentalism and liberal democracy.

"Quants" on Wall Street

Recent years have witnessed an influx of people with scientific training into Wall Street.[4] They arrived as a result of new developments in the financial world but their arrival in its turn changed that world in a very substantial way. Why was Wall Street so receptive to such people? What did they use them for? "Quants," as they are called, bring with them a particular way of thinking and of approaching problems; they bring with them the culture of science. They are comfortable with sophisticated mathematical models and algorithms and sometimes produce their own by making analogies between the worlds of science and of finance. What is the practical value of such models and algorithms? What influence does the existence of such models have on the financial system? In large part, this kind of math is used to manage risk, which is the name of the game in banks, brokerage houses, hedge funds, and insurance companies.

It is not coincidental that Wall Street has found an important niche for people with mathematical and scientific training. I remember when a friend who ran the pension fund for a large corporation suggested I go into the consulting business because I was a mathematician. He wasn't so much interested in my particular expertise as he was in the impression that such expertise creates in the uninitiated. He even had a proposed name for the hypothetical company: "AlgorithmsRUs." In the uncertain world of trading stocks and bonds, nothing sells so well as the "sure thing." Algorithms, mathematical formulas, and equations provide the promise of certainty. The aura that science provides—precision and objective truth—migrates over into the field of finance. However, economics and finance cannot realistically expect to have the exactitude of the physical sciences. If the claims of absolute certainty in physics and mathematics can be disputed (as we shall discuss in depth in subsequent chapters), how much more so can these claims be disputed in "softer" disciplines, which deal with human behavior.

The sad economic history of the recent past has a lot to teach us about science and its effect on society. What caused bankers who at one time were ultra-conservative and adverse to risk to become such gamblers and risk-takers? You could suggest that the explanation for this changed behavior is very simple—greed. But there is something deeper going on. Why were the stakes raised continually until eventually the economy of the entire world was at risk? The answer is that people convinced themselves that the risks were not that great, that these new mathematical tools would ensure that money could be made no matter whether the market went up or down.[5] This is the same kind of thinking that is the downfall of every person who goes to Las Vegas with a "system." The financial people also had a system, one that only a few "quants" could understand. The complex mathematical instruments that they supplied promised to convert the seeming randomness of the stock market into something that was orderly and predictable.

It was easy to conclude that this kind of thinking could make you rich, and in many cases, it did. If certainty in the financial world is unobtainable, it is still possible to package the illusion of certainty and use it to make money. The package consists of precisely those algorithms and equations, not science but the mythology of science, not certainty but pseudo-certainty. The more complex the package and the more arcane the mathematics, the better. What was being sold was the faith that the complex, *human*, world of economics and finance could be made over in the image of science, could be made objective and predictable. It really didn't matter whether the CEOs of these companies understood the math and, in many cases, they clearly did not. The less it was understood, the more powerful it was as mythology—a screen against which human desires could be projected.

I don't mean to say that financial mathematics has no value, only that it is not infallible. Science works well for a certain range of phenomena but tends to break down when you get to the limits of its natural range. At a certain point these financial instruments broke down with disastrous results. The entire process revolves around the human need for certainty, the notion that science produces certainty, and therefore that the same kind of thinking will produce the same certainty in domains like economics.

Pseudo-Objectivity through Quantification

Modern society is well on its way to realizing a dream that dates back to the Pythagoreans and their numerical mysticism. This is the dream of the "world as number." Today we are busy quantifying the world, reducing it to number. The evidence is all around us in the computers we use and the digital technology through which we listen to our music or watch our favorite movies. The world has gone digital in a big way, but we have not yet uncovered all the implications of this revolution.

Let's begin with a simple example of quantification. Suppose you want to encourage good teaching. You would first have to evaluate it, and you might do so by giving students a questionnaire that measures different aspects of teaching: does the teacher arrive on time, does he understand the material, does he speak well, and so on. Give each one of these a number from 1 to 10 and average the numbers. This gives a numerical measure of the teaching, and you might say that a person with an average of 8.5 is a good teacher and one with 5.3 is a bad one. This makes things very easy for the person doing the evaluation because there now appears to be an objective criterion for establishing the quality of teaching. Similar procedures can be found in every human relations department in any company or institution. The drive for certainty is what is behind such numerical reductionism. Numbers make some kinds of decisions easy, but all too often this kind of thinking is a simplistic cop-out. It reduces a judgment, whose essence is qualitative and subtle, into one that is quantitative and easy. In our example, you have a number, but what does that number reveal? Can a teacher who wishes to attain a better number do so by manipulating the situation without actually becoming a better teacher? Will this number pick out those exceptional and devoted teachers who manage to inspire students and change their lives? Of course not! This kind of quantitative reductionism is one of the main elements behind the financial crisis. People have allowed complex, mathematical-based formulas to obscure human intuition and judgment.

This tendency toward quantitative reductionism is the subject of an interesting book, *Trust in Numbers: The Pursuit of Objectivity in Science and Public Life*, by historian Theodore M. Porter.[6] He discusses the history of the pursuit of a kind of mechanical objectivity through numbers:

> By now, a vast array of quantitative methods is available to scientists, scholars, managers, and bureaucrats. These have become extraordinarily flexible, so that almost any issue can be formulated in this language. Once put in place, they permit reasoning to become more uniform, and in this sense more rigorous. Even at their weakest point—the contact between numbers and the world—the methods of measurement and counting are often either highly rule-bound or officially sanctioned. Rival methods are thus placed at a great disadvantage. The methods of processing and analyzing numerical information are now well developed and sometimes almost completely explicit. Once the numbers are in hand, results can often be generated by mechanical methods.

What we fail to see is that the process of quantification in its attempt to "capture" a given situation actually modifies that situation. It is the classical effect of self-reference that I shall discuss in chapter 6. *Putting numbers into a situation changes the situation.* This element of self-reference usually goes unnoticed and therefore is not accounted for. Quantification does not "capture" reality as much as it creates a new reality.

Mathematics is a powerful means to win arguments because people have a strong feeling that mathematics is objective, that "figures cannot lie." The truth is that it is easy to mislead and obfuscate a situation through the use of mathematical and statistical models that are inappropriate, whose assumptions are simplistic or just wrong. Politicians make use of mathematics in this way. I remember a recent interview with the Prime Minister of Canada after a G8 summit meeting. Stephen Harper was arguing that the developing world would have to control their emissions of green house gases. In his attempt to move the onus for change away from the industrialized world in general, and Canada in particular, he made the comment, "So, when we say we need participation by developing countries, this is not a philosophical position. This is a mathematical certainty." Of course, Harper never showed reporters the assumptions on which his statement was based. It sufficed for his purposes to put together the words "mathematics" and "certainty." We tend to believe that the best decisions can only be made in a context of complete certainty about the parameters of the situation and we believe that science and mathematics can, indeed must, provide us with this certainty. In this way, science and certainty are inextricably linked.

Black Swans and the Misuse of Statistical Models

Nassim Taleb has written a couple of best-selling books in which he makes a passionate case against the kind of simplistic statistical models that he feels contributed mightily to the current financial crisis. He also discusses the cultural and philosophical background of these events in terms that have a great deal in common with the point of view I take in this book. He asks, "Are we using models of uncertainty to produce certainties?"[7] The answer is obviously "yes, we are," and that is why we are in the present predicament. But he says that the problem does not lie with professional statisticians who "can be remarkably introspective and self-critical." The problem lies with the "commoditized, 'me-too' users of the [statistical] products.... Those in other fields who pick up statistical tools from textbooks without really understanding them.... Alas, this category of users includes regulators and risk managers, whom I accuse of creating more risk than they reduce."

Taleb has gained a large following because he was so prescient in predicting the problems that would subsequently arise in many large American financial institutions. He makes a very convincing argument about the dangers of the use of simplistic (even if highly technical) mathematical models.

The manner in which we attempt to control risk increases the probability of catastrophic events. This shows us that there is something fundamentally wrong. It is not just that people do not understand the math and apply it inappropriately. If that were so, then we could just produce a better model. The problem is that the new model will have its own black swan (highly improbable) events, that uncertainty is so intrinsic to the situation that it will inevitably appear. In fact, the odds of it appearing are compounded by the techniques created to avoid it and the mental set that does not allow for uncertainty.

Taleb makes the point that black swan events are only predictable *after the fact*. Beforehand, they are unpredictable and cannot even be understood. After-the-fact explanations are useful, but they are like rationalizations: They explain away the crisis and therefore make us feel better. However, they do not change the inevitability of such "blind spots," the element of intrinsic randomness and unpredictability that is the backdrop to any realistic description of the world.

Organized Religion

Scientific theories are not the only ones with the potential to provide certainty. All cultures are supported by a mythological structure, systems of beliefs that give meaning to people's lives and so support the difficult business of living. An even older system of beliefs comes from organized religion. In the struggle for dominance between religion and science, the balance of power and influence has slowly but steadily swung in the direction of science and technology. Yet both science and organized religion retain a significant presence in the contemporary world. The current version of this struggle could be thought of as the battle between two fundamentalisms: the scientific and the religious. The former is in love with technology and has a deep faith in the power of science and rational thought to solve the most intractable problems. It has little place for things like meaning and morality—it sees life as the result of impersonal, random forces. It is exciting and dynamic but can also be frightening and alienating. Religion carries with it a strong sense of meaning and a particular set of moral values—a sense of the purpose of human life. Yet it often has little tolerance for those who do not share its particular beliefs.

The struggle between these two cultures threatens to destabilize the world and perhaps may even destroy us all since even fundamentalists can build nuclear and biological weapons. The problem is that these two certainties are seemingly incompatible with one another. Most people who read this book will be members of the scientific culture and will object to my seeming to draw a parallel between these two cultures. Recently, there have been a number of vastly popular books by people like biologist Richard Dawkins and journalist Christopher Hitchens that paint organized religion as a great evil. Of course, millions of devout Christians, Jews, Muslims, Hindus, and Buddhists often see things in exactly the opposite sense—the evil, for them, lies in the scientifically based world of modernity. In the middle are many who want the best of both worlds, who want the security of a well-established system of values and beliefs as well as the material benefits of the consumer society that arises from the innovations of science and technology. This latter group notwithstanding, it seems clear that the certainty of conventional religious beliefs is increasingly incompatible with the certainties of the world of science. This incompatibility is part of the social and cultural crisis that the world finds itself in today.

This book is not about religion and not even about the argument between science and religion. It is about mathematics and science, but comparing science to religion for a moment helps me reinforce the argument of the earlier chapters about the existence of two divergent tendencies within science.

It is the science of certainty that is engaged in a battle to the death with the religion of certainty. The science of wonder is perfectly compatible with a religion of wonder, for ultimately they are the same thing.

The religion of certainty will be fairly clear to most people—organized religion, for the most part, fits into this category; the religion of wonder is less familiar. What differentiates them is the role of certainty. What is the difference? The Russian writer Fyodor Dostoevsky made the crucial point many years ago in his greatest novel, *The Brothers Karamazov*. There is a famous section of the book called "The Grand Inquisitor" where Jesus Christ visits the Spain of the Spanish Inquisition—a time when thousands of so-called heretics were burned at the stake, condemned to death by religious authorities. In the novel, Jesus quickly runs afoul of these same authorities and is jailed as a dangerous radical and a threat to the established order. In a famous scene, the Grand Inquisitor confronts the prophet reborn. The inquisitor's self-justifying argument is a familiar one: It places human beings' need for certainty against their need for freedom. Freedom, the Inquisitor says, is a threat to the power of organized religion, to the stability of society and thus to the public good. People and society, he says, need rules, order, and stability. Freedom is a luxury they cannot afford. Prophets are best restricted to the distant past. In the present, they are at best an inconvenience and at worst a threat. His argument is that people just cannot cope with uncertainty; they cannot manage freedom—"for their own good" they need the imposition of certainty.

In the history of religion, it is easy to see how the extraordinary creative insights of the founders of these religions have, in many cases, hardened into rigid, authoritarian ideologies. As a result, most people have trouble juxtaposing the words religion and creativity. The progression from the freedom, flexibility, and openness that are properties of creativity, to the rigidity of fixed and certain systems of thought seems almost inevitable because we have seen it happen so many times in history. The same progression can be found in the history of science, most notably in instances of the kinds of "paradigm changes" that have been identified by the well-known philosopher of science, Thomas Kuhn.[8] Science is a domain of human activity, and so the same rules apply to science as they do to other human activities. Science and mathematics have their own unique version of this primordial argument between the need for certainty and the need for creativity, between ideology and freedom. Mathematics is, in a way, the perfect exemplar of a "science of certainty" and yet mathematics is also a domain of wonder and the deepest creativity. In science and mathematics, we shall find a unique way of approaching certainty and creativity that will also tell us something about the general relationship between our conflicting needs for security and freedom.

We all claim that we prize freedom but freedom is not such an easy thing to live with. It has been said that the price of freedom is eternal vigilance. We must always be vigilant about the tendency to trade in our freedom for the illusion of security and certainty. The danger is ideology, the power of an idea or a set of ideas. Science and religion both have the potential to become ideologies with potentially disastrous consequences. The idea that science inevitably will save us from ideology is itself an ideology. Science, like any other human activity, carries this tension between certainty and wonder. Science does great damage when it turns into ideology, when it begins to worship certainty.

5

‿❧‿

Ambiguity

*We must learn to be comfortable with not knowing, with
ambiguity and uncertainty.*
—Arthur Zajonc[1]

INTRODUCTION

To live in a world of absolute certainty, even if it is only the dream of certainty, is not wrong but it is limited since it involves looking at only one side of the total situation. Acknowledging the existence of a blind spot implies a willingness to live with the other side, the side that contains uncertainty and incompleteness as well as ambiguity and even paradox. In this chapter, I focus on the phenomenon of ambiguity. The usual take on ambiguity is that it reflects a break down in certainty and is "corrected" by removing the ambiguity, thereby reestablishing certainty. The ambiguity is not seen as having any value in its own right. However, in this chapter and those that follow, we shall see that ambiguity cannot be permanently avoided, and, moreover, that it has value in its own right. If we take seriously the existence of a blind spot, then we might ask about the existence of a mechanism for melting down the rigidity of a state of permanent certainty. Ambiguity, as we shall see in the following, is that mechanism. It contains two points of view (like certainty and wonder) that are both valid but which conflict with one another. It is this conflict that forces us beyond either partial viewpoint.

Ambiguity represents a new approach to science, an approach that is open, and flexible; an approach that has room for the problematic and the creative. As we investigate ambiguity in science in the next two chapters, we shall see that it is a way of evoking the blind spot. Ambiguity contains a blind spot similar to those problematic scientific concepts we discussed in chapter 2: zero, infinity, and randomness.

What place does ambiguity have in science? In 2007, Gaurav Suri and Hartosh Singh Bal published a novel about mathematics called *A Certain Ambiguity*.[2] Its theme is certainty—the human need for certainty, the attempt to find certainty through mathematics and, ultimately, the need to confront uncertainty. One of their characters says, "Human beings can never be certain about anything, and without certainty there cannot be meaning,"[3] which is a good statement about the dilemma of modernity. But let us go back to the title for a moment, "a certain ambiguity." The title is itself ambiguous. It could be referring to one particular ambiguity among many but, more likely, it means that ambiguity is certain, that it is inevitable. You could even turn that last phrase around and get yet another possible meaning—that certainty is ambiguous. Thus, both the title and the mathematical certainty that the book discusses are ambiguous, and that on a number of levels.

The title juxtaposes certainty and ambiguity, a most unlikely combination. At first glance, certainty appears to be the opposite of ambiguity. Science and mathematics, it would seem, are means through which to establish certainty and eliminate ambiguity. However, I propose to investigate "science in the light of ambiguity." Of course, this phrase is also ambiguous. The usual phrase is 'the light of reason." The light of reason refers to what I called the science of certainty, the light of ambiguity to the science of wonder and creativity. Just as it is unusual to base a description of science on characteristics like mystery, wonder, and creativity, so it is equally peculiar to discuss ambiguity in the context of science. Yet this is what this chapter is about. The discussion of certainty and uncertainty leads naturally to a discussion of ambiguity. Normally, one imagines that certainty is the way things are, and ambiguity is a preliminary stage that arises before we have managed to sort things out. I agree with David Bohm, who said, "the ambiguous is the reality and the unambiguous is merely a special case of it, where we finally manage to pin down some very special aspect." Ambiguity is the way things are. Certainty is temporary; it comes and goes.

THE AMBIGUOUS NOTION OF AMBIGUITY

If you look up ambiguity in the dictionary, you will find two meanings. The first is "vague or unclear"; the second relates to having a double meaning and comes from the prefix "ambi-" such as one finds in the words ambidextrous or ambivalent. For the moment, I shall focus on the second meaning, although it is amusing to note that the word ambiguity is itself ambiguous according to the second of these two meanings. The following is a formulation that will help us start thinking about ambiguity.

> *Ambiguity involves a single situation or idea that is perceived from two self-consistent but habitually incompatible frames of reference.*[4]

When there is ambiguity, there is one situation but two perfectly good ways of looking at it. To make matters worse, these two points of view are in conflict and may even be incompatible with each other. This incompatibility makes situations of ambiguity uncomfortable, irritating, even anxiety provoking—it evokes a tension that might even feel intolerable.

Generally speaking, the demands of ambiguity have two possible responses. One can eliminate the conflict by asserting that one point of view is correct and the other incorrect. The second possible response involves an act of creativity. Here the incompatibility is viewed as a challenge or an opportunity for a deeper understanding of the given situation. The tension is viewed as a friend to be cultivated, not as an enemy to be eliminated. The ambiguity is then the very crucible of creativity. If you want to make major progress in science, you must be prepared to pay the price. The price involves staying with the incompatibility and the resulting tension, and living with it for extended periods of time. Isaac Newton described the secret of his successes in the following words, "It was through concentration and sheer dedication. I keep the subject constantly before me, till the first dawning opens slowly, little by little into the full and clear light."[5] The element of frustration tolerance is not usually the factor that is focused on when scientific breakthroughs are discussed, yet it, not raw intellectual power, may be the essential factor in acts of creativity. This gives another way of thinking about Thomas Alva Edison's saying, "Genius is 99 percent perspiration and 1 percent inspiration"—the perspiration involves staying with the tension until that magical moment when inspiration reveals itself.

In such cases of ambiguity, the problem is resolved by moving to a higher point of view from which both of the original perspectives have value. The new viewpoint reconciles the conflict; the original perspectives are now seen as two aspects of something more general.

One metaphor for ambiguity is binocular vision. When you cover up one eye and view a scene through the other, the scene you see is flat, two-dimensional. When you look at the same scene with two eyes, each eye registers a slightly different scene. These are the incompatible points of view. The brain reconciles these two views by creating a new way to see the situation. This is accomplished by introducing a new dimension—depth. Without the initial incompatibility there would be no depth. This metaphor brings out a number of important points and serves to distinguish the "ambiguous" point of view from the strictly logical. The first is that the incompatibility that exists in a situation of ambiguity *has a potential value* that would be lost were we to prematurely resolve the conflict by arbitrarily eliminating one or the other initial viewpoint. Logical incompatibilities are called contradictions, and the idea that one should work with these situations rather than eliminating them has profound implications. The second is something that every scientist understands: that one of the things sought in science is depth, as opposed to superficiality. What makes one idea profound and another trivial? The answer to this question will not be found in the logical elements of a scientific theory but may well be understood via the notion of ambiguity.

Ambiguity is not something that is static in time; it is something that happens. The very conflict that appears irresolvable at a given moment may open up into a novel landscape within which one operates with a new openness and flexibility. Thus, ambiguity is tied to the processes of learning and understanding. It is tied to moments of insight, what has been called the "Eureka" or "Aha!" experience. Ambiguity reveals the dynamics of the creative breakthrough in science but, more prosaically, it also illuminates ordinary acts of learning and understanding that are experienced by everyone. Before the insight, the ambiguous situation appears to be incomprehensible. Afterward, it is simple and open. These conflicting frameworks now give rise to a new freedom to go back and forth freely from one point of view to the other. Thus, ambiguity is dynamic and tied to the creative process.

The kind of situation I am referring to by using the term ambiguity is not at all unusual. In the arts, reference to ambiguity is very common. Leonard Bernstein points to ambiguity as the key to expressivity in music, "the more ambiguous the more expressive."[6] One well-known book about literature is called *Seven Forms of Ambiguity*.[7] As far as literature is concerned, every metaphor is an example of ambiguity. This would make ambiguity the building block of poetry. It is more unusual to discuss ambiguity in the sciences, as I have done, but ambiguity is as fundamental to the sciences as it is to the arts. However, if this is indeed the case, it will necessitate rethinking the nature of science and mathematics. At the most obvious level, ambiguity is precisely the quality that distinguishes the arts from the sciences. The arts admit ambiguity as legitimate; the sciences do not. The story that mathematics and the sciences tell about themselves—the meta-narrative if you will—is that logic, mathematics, and the sciences overall stand as a bulwark against sloppy thinking in general and ambiguity in particular. In my analysis, this description might well hold for what I called the science of certainty, but not for the science of wonder.

The characterization of ambiguity that I expressed earlier is very similar to a description that Arthur Koestler gave some years ago of the creative process.[8] His description of creativity involves replacing the word "ambiguity" in the earlier description with the word "creativity." This reinforces my point that ambiguity is intimately tied to creativity. In fact, the psychologist John Kounios defines creativity as the "ability to restructure one's understanding of a situation in a nonobvious way."[9] In a situation of ambiguity, such a restructuring is evoked by the need to resolve the conflict between incompatible frames of reference.

The author Albert Low is the person who has most forcefully and persistently pointed to the importance of ambiguity in the ecology of thought.[10] Low's usage of the term emphasizes the generative nature of ambiguity, and I shall also use "ambiguity" in this way. Ambiguity is not so much a concept as a situation that produces concepts. Thus, when one uses the term ambiguity, one is not talking about a static and objective situation—something that is "out there." One is talking about a situation that is dynamic, that has the capacity for change.

When we come up with a description of reality that uses language and concepts, it is as though we are painting a picture of reality. It is clear that the picture is not reality—"the map is not the territory," as they say. Science is inevitably such a map. Ambiguity, as the term is used in this book, is an attempt to break down this formidable barrier. It is an attempt to come up with a process description of science using words that are appropriate to the task. This is why I am hesitant to say that ambiguity is a concept. This would stabilize it—give it a fixed meaning. In chapter 8, I discuss why even a very basic term such as "number" has no precisely defined meaning that can be fixed for all time. Ambiguity is a tool that is being used in an attempt to develop a description of the scientific world without becoming fixated on certainty—that is, without demanding an unchanging objectivity. It would be a little inconsistent if I were to insist that ambiguity itself has a fixed, objective meaning. What I will do to avoid this trap is give ambiguity the preliminary definition, used earlier in these pages, which works well enough to begin the topic, and then later expand this initial approach.

Low often describes ambiguity by referring to the famous gestalt picture of the young woman and the old lady.[11] (See figure 2.) This picture has much to teach us about the nature of ambiguity. It could be taken to be an abstract field of black and white dots, but of course this field has two viable and consistent interpretations: the old lady and the young woman. Notice that each of these interpretations describes the entire field—nothing is left out; everything is accounted for. Yet the two interpretations are in conflict with one another. You cannot see the young woman and the old lady at the same time. What happens at the beginning is that you may see only one view—for example, the picture of a young woman. This situation is now stable and "objective," but this objectivity is obtained at the cost of eliminating the other viewpoint. Suppose at a certain moment you see the second view—the old lady, say. At this point, the situation loses its stability since you will now tend to alternate between the two views.

One might ask, "What is the picture objectively?" However, this question has no meaning. The problem with this picture is that it is not static. It won't stay still, so to speak. The only way to keep it still is to look at it

Figure 2

as a field of black and white dots. But if you do so you are rejecting the ambiguity and thus missing the essence of the situation.

It is interesting that Thomas Kuhn used a similar gestalt picture, the duck–rabbit illusion, to explain what he meant by his famous idea of the paradigm-shift in science.[12] The shift from one paradigm to another is "a relatively sudden and unstructured event like the gestalt switch." This is why Kuhn calls different scientific paradigms "incommensurable" with one another. The word incommensurable, a strong form of incompat-ibility, is taken from the Greek discovery that irrational numbers, like the square root of two, cannot be expressed as the quotient of two integers and therefore have no common measure with rational numbers. Incommensurability is Kuhn's way of highlighting the radical conflict that exists between the points of view represented by different scientific paradigms. He says, "because it is a transition between incommensurables, the transition between competing paradigms cannot be made a step at a time, forced by logic and neutral experience. Like the gestalt switch it must be made all at once or not at all." What Kuhn calls "revolutionary science," the situation in which the old paradigm has lost its power, even though its successor is not yet established, is evidently a situation of ambiguity. Each paradigm represents a consistent framework, yet the two frameworks are incompatible with one another.

Now, if you feel that the gestalt picture and ambiguity in general are exceptional and needlessly complicated I would suggest we take another look at some basic ideas in mathematics and science. When we come to see mathematics and science as fundamentally ambiguous, we shall feel the earth moving under our feet, so to speak—we shall be living in a new paradigm.

Ambiguity in Arithmetic

How can mathematics, the most logically rigorous of disciplines contain ambiguity? Let us begin by considering simple arithmetic. Take the multiplication of whole numbers—3×2, for example. Multiplication is usually introduced to students as repeated addition. Thus, 3×2 means three 2's or $2 + 2 + 2$. This reduces multiplication to addition and is the normal way in which new operations are introduced—by building on old ideas or skills. Students who have strong skills in mental addition tend to reinterpret every question about multiplication as a question about addition. However, under the pressure that arises from more complicated problems, like 37×56, this way of looking at things quickly breaks down. Here, reducing things to repeated addition is not very useful. The student who has strong computational skills in addition may, *for this very reason*, be the one who fails to make the conceptual leap to multiplication as an independent operation in its own right.[13] Success in mathematics is not based wholly on previously acquired skills or some logical or abstract "intelligence." Intelligence may well involve the ability to cope with situations of disequilibrium in the learning process such as moving from thinking of multiplication as repeated addition to thinking about multiplication in a new way.

Consider the commutative law of multiplication, that $3 \times 2 = 2 \times 3$. Why is this valid? For the child armed with only the first definition, it is not immediately clear why two 3's gives the same result as three 2's. From the point of view of repeated addition, you could look at this result by observing that $3 \times 2 = 6 = 2 \times 3$. This might seem obvious, but to understand this statement the child needs to perform some thinking that is quite subtle. We introduced 3×2 as an operation or process. When we assert that $3 \times 2 = 6$, we are equating this process to the *result* of this process. We are saying that an action is equivalent to an object, namely the results of that action. In order to fully appreciate the subtlety of the equation $3 \times 2 = 6$, we must become comfortable with this ambiguity: multiplication can be thought about in two ways, as operation and as number. As object, 3×2 is a number. You don't have to perform the operation to assert it is a number. For example, even if you are multiplying two enormously large numbers, such as $1234567890 \times 987654321$ one can assert *without any calculation* that the product is a number. If the product, whatever its value, were to be divided by 987654321, then the answer would of course be 1234567890. Thus, multiplication is ambiguous, both process and object, even though one could well claim that these two ways of looking at multiplication are really quite different, even that they are in conflict. This does not prevent us from thinking of 3×2 as one (ambiguous) idea. The well-known mathematician William Thurston makes much the same point when he discusses that momentous insight (for him, as a child) of discovering that 134/29 is a number and not just a problem in long division.[14]

Simple multiplication is ambiguous in more than one way. For example, 3×2 is often understood as the area of a rectangle with sides of length 3 and 2. Thus, there would be two ways of understanding 3×2, as repeated addition and as area. These are surely non-equivalent (but related) representations. The child must come to see 3×2 as the one concept that stands behind these two representations and master *this* ambiguity. Ultimately what is required is a certain mental flexibility that comes with seeing that the many different ways of thinking about multiplication are different ways of thinking about the same thing. Thus, one idea—multiplication—that can be understood in a variety of ways, and one is free to move from one of these points of view to another. We sometimes think that because we use the same word—in this case, multiplication—we are always talking about the same thing. But even though we use the same symbol for multiplication, the multiplication that is symbolized by "×" in 3×2 is not the same multiplication in -3×-2 and even less, the multiplication in $\pi \times \sqrt{2}$.

When one expands the number system to negative numbers, the situation becomes even more complex. In the first place, a negative number is intrinsically ambiguous. "-3" can be thought of as either a number or the process of subtracting three units. When one then multiplies such numbers and writes down the equation that is so mysterious to almost all students—namely, $(-3) \times (-2) = +6$—the immediate puzzle is why you use the same symbol for multiplication in these two situations (3×2 and -3×-2). If multiplication is identified with area, then the multiplication of negative numbers does not seem to make sense. Multiplication must be freed from its identification with this specific representation in order to set the stage for the successful leap to the higher-level concept involved in the multiplication of negative numbers.

The mathematics educators Eddie Gray and David Tall[15] coined a name for these types of situations where something is simultaneously a process and a concept. They called this ambiguous animal a "procept" and pointed out that such situations pervade all of elementary arithmetic. A procept is ambiguous because it involves two frames of reference: process and object. Yet the fact that a procept is unitary—that it is one idea—is brought out by the fact that this ambiguous situation is represented in mathematics by a single symbol. 3×2 is both the number and the operation; $3/2$ is both a fraction and the number that results from the operation of division; -3 is both number and subtraction, and so on. Learning involves leaving behind the identification of these procepts with one specific frame of reference and coming to see them as the ambiguous creatures they really are.

Simple arithmetic teaches us a great deal about the nature of ambiguity. It shows us that ambiguity is itself something that has a dynamic and not a static nature. This is why ambiguity is associated with creativity, and learning is a form of creativity. An ambiguous situation does not just sit there passively like some formal theory. It always contains a potential leap of insight. In this book when I say that a given idea or situation is ambiguous, I will often be referring to its dynamic nature as well. If we think about science as being made up of conceptual objects with monolithic meanings that are fixed and unchanging, then we have a science that is essentially static, but if we think about scientific concepts as intrinsically ambiguous, we have a science that not only describes change but that itself contains an intrinsic potential for change.

It turns out that not only is arithmetic founded on ambiguity but also much of the rest of mathematics. When one writes down an infinite decimal, $1 = 0.9999\ldots$, then one is clearly referring to the number 1 as both object (number) and process (the sum of the infinite series $9/10 + 9/100 + 9/1000 + \ldots$). This ambiguity is written into the notation for infinite series. It is the cause of students' difficulties with the concept of a real number, yet it is a triumph of the invention of the real number system. Thus, ambiguity is both a problem—the source of difficulty—and the solution to the problem; the resolution of the difficulty in an act of creative understanding.

Ambiguity in Algebra

Much of science gets its effectiveness from the mathematical notation in which it is written. The most basic aspect of mathematical notation involves the use of algebraic methods, and this means the use of variables, the ubiquitous "x" in expressions like $3x + 7$, in equations like $3x^2 + 2x + 1 = 5$, in functions like $f(x) = \sin x$, and so on. The introduction of algebraic notation, of variables, constituted a huge leap in the development of scientific thought. Variables convey great power to those who learn how to use them correctly, but where does that power come from? The power of algebraic notation stems from harnessing ambiguity! What do I mean by this?

Think of the expression $3x + 7$. What does the "x" stand for? Well it stands for any one of a whole set of numbers. Let us restrict our discussion to the counting numbers $N = \{1, 2, 3,\ldots\}$. So "x" stands for any one of these numbers or else the whole set simultaneously. In the latter case, $3x + 7$ would represent the set $\{10, 13, 16, 19,\ldots\}$. Here we already have the appearance of ambiguity: Does $3x + 7$ represent *one* of the possible values or *all* of them. The answer is *both* and *neither*. When a student works with such an algebraic expression, she thinks of "x" as some un-specified but specific counting number, but she doesn't know what it is. Even though you don't know it, you work with it as though you did. For example, to solve the equation $3x + 7 = 13$, you imagine that you know "x," and since $3x + 7 = 13$, then $3x$ must equal 6 and "x" must equal 2. At each stage in this simple derivation, you both know "x" and don't know it. At the beginning, "x" could be any number; at the end, it can only be 2. But at the beginning it is still implicitly 2. And at the end you are simultaneously saying that any number $x \neq 2$ is not a solution. So at every stage the "x" is ambiguous—it stands both for every number in its domain, as well as some specific number. Working with variables means carrying this ambiguity with you from the first line of the computation until the resolution where the ambiguity seems to disappear in favor of the answer. However, in reality the ambiguity never really disappears. It is harnessed toward definite ends.

It is this ambiguity between the general set of possible numbers and the specific but unspecified number that makes algebraic notation work. The essence of algebraic notion, the essence of variable, lies precisely in its ambiguity. This is the reason why algebra is difficult for some children to learn. The difficulty lies precisely in the ambiguity. Of course, the value of algebraic technique also lies in the same ambiguity.

Although the notion of variable is brought out most clearly and force-fully within the context of mathematics, what is going on here is per-fectly general. It arises anytime we think of some object as a specific but unspecified member of a class. The word "tree" is an example of a vari-able. When we use the word tree, we think of the generic tree—in other words, of any tree—but the word may simultaneously evoke a picture of some specific tree. We think and talk simultaneously in the concrete and the abstract. This is the power of language just as it is of mathematics.

In discussing the ambiguous notion of "variable," we have revealed a very deep and very general characteristic of thinking, of how the mind is used in language and in science. The natural world is unwieldy. It con-sists of a seemingly infinite number of specific instances. Everything is unique and incomparable! How then can we say anything intelligible about such enormous collections of objects? In mathematics, this is the question of how the infinite can be reduced to the finite. Mathematics has been called the science of the infinite—every mathematical theorem refers to an infinite number of cases, "the sum of two odd numbers is even," for example. Yet mathematics makes intelligible statements about such infinite situations. All of science involves the reduction of the many to the one, so to speak. The mechanism through which this is accomplished is the systematic use of ambiguity such as we saw in the case of variables in algebra.

Ambiguity in the Theorems of Mathematics

Even if it is true that mathematics and science have discovered a way of using ambiguity in a controlled manner in their notation and procedures, surely the results of science and mathematics—the theorems and scientific laws—are unambiguous. Shockingly this is not the case at all! In the next example, we will discuss a famous theorem whose importance lies precisely in its ambiguity.

Calculus comes in two varieties: differential and integral calculus. They have distinct histories. Integral calculus is a generalization of the study of area and was essentially invented by the ancient Greeks, most notably Archimedes. Differential calculus arises from the study of the rate of change of one variable with respect to another. Thus, it might concern itself with calculating velocity or acceleration or rates of exponential decay. Geometrically this amounts to calculating the slope of tangent lines to curves. The development of this theory was originally due to Newton and Leibniz. Thus, we have two calculus disciplines that arise from the consideration of different problems and, at first glance, have little to do with one another.

The fundamental theorem of calculus says that these processes—differentiation and integration—are the inverses of one another. This means that differentiation cancels out integration (in a certain specific sense) and vice versa. But the deeper meaning of this theorem is that calculus is ambiguous—there is one calculus with two frames of reference: the usual differential calculus and integral calculus. This does not mean integral calculus is identical to differential calculus. It does not mean we can now dispense with differential calculus, say, and make do with only integral calculus. It means that, like the picture of the young woman and the old lady, we now have two ways of looking at calculus. We have two languages, so to speak, and a way—the fundamental theorem—of translating back and forth between them. This is a great advantage, for there are now all kinds of things we can do with our new multiple point of view that we could not do without it. For one thing, differentiation is relatively easy but integration is hard, so we can integrate by using the rules that we have established for differentiation.[16]

The essence of the situation is that calculus is ambiguous and it is the recognition of this ambiguity that is the great creative breakthrough captured by the fundamental theorem. The essence of many advances in mathematics lies precisely in the recognition of "one idea that may be viewed from multiple perspectives." It is, for example, the essence of the Taniyama-Shimura-Weil conjecture, a key element in the recent resolution of Fermat's Last Theorem by the mathematician Andrew Wiles.[17]

Ambiguity in Classical Logic

The naïve approach to science would have it that science is merely the systematic investigation of the natural world. Many people feel that science involves looking at things as they are in reality, and so would resist the suggestion that science is built upon a framework of assumptions—that science is a culture. Science is founded upon many arbitrary, often unexamined, hypotheses. A major aim of this book is to reveal these hidden assumptions and to undertake a critical examination of them. One of these assumptions involves the role of reason, more specifically the role of classical logic. The whole discourse of science is built upon the assumption that one can draw correct inferences by using the methodology of logic.

Logic itself is never questioned. The exception being a small group of physicists who felt that quantum mechanics revealed the need for a more subtle logic. Nevertheless one aspect of classical logic that no one would question is the idea that it is unambiguous. In fact the whole point of logical discourse is that it dispels ambiguity and enforces clarity. Since it is usually believed that the universe itself is clear and unambiguous, logic would not be the imposition of some arbitrary feature on the natural world but a reflection of the "way things are." Many philosophers even believe that classical logic is built into the structure of the universe. They might not put it that way, but they feel that logic is the way things are and they are clear and unambiguous.

In opposition to all such beliefs, it is my contention that classical logic is itself ambiguous in a very obvious way. This brings into question the relationship between logic and the natural world, not to mention the relationship between logic and science. So how can I say that logic is ambiguous?

The discussion revolves around the notion of logical equivalence, which is called "tautology." A tautology consists of two statements that have the identical (logical) truth-value. If we denote the statements by **P** and **Q**, then a tautology would have the form "if **P** is true, then **Q** is true," and simultaneously "if **Q** is true, then **P** is true." We also say "**P** if and only if **Q**" or "**P** is a necessary and sufficient condition for **Q**" and write $\mathbf{P} \Leftrightarrow \mathbf{Q}$.

The normal assumption is that tautology means identity, that two tautological statements are essentially identical. For example, there is the famous law (tautology) of logic called "**modus tollens**," which asserts that the implication "**P** implies **Q**" is the same as the implication "**not Q** implies **not P**." Is this tautology "self-evident" or "obvious" in the sense that it is implicit in the very meaning of the words themselves or is it itself an assumption? Is a tautology necessarily true or is it possible to draw a distinction between truth and tautology, between truth and logical truth?

Tautologies are not self-evident truths built into the fabric of reality itself. On the contrary, the essence of a tautology is precisely that it *is* ambiguous! Two examples from elementary mathematics will make this point crystal clear. First, consider the (true) mathematical proposition, "if the square of some integer is odd, then the integer itself must be odd." By **modus tollens**, this is tautologically equivalent to "if the integer is even, then so is its square." Strangely enough, when we come to attempt the verification of these two propositions, we discover it is very difficult to find a direct proof of the former, whereas the latter is almost immediate. (If n is even, then n = 2k for some integer k. Then, $n^2 = 4k^2 = 2(2k^2)$, and so n^2 is even.) The method of proving the first statement by means of transforming it into the second is a method of argument that goes by the name "proof by contrapositive." Even though the two statements are logically identical in one way, at the mathematical level they are subtly different.

The second example also involves a tautology. This one states that a number, x, is rational (a fraction) if and only if its decimal representation is finite or repeating. Thus, $1/3 = 0.3333\ldots$, $1/4 = 0.25$, $0.125 = 125/1000$, and $0.121212\ldots = 12/99$. Normally, saying that a number is rational amounts to saying that it can be written as the quotient of two integers, $x = m/n$. So saying that it is irrational amounts to saying that it can *not* be written in this way. It gives no information about whether such irrational numbers exist. (The Greeks discovered a way of showing that geometric numbers, such as the square root of two, were irrational using this negative criterion.) However, saying that the decimal representation is *not* terminating or repeating gives us a simple criterion for determining which real numbers are irrationals and for distinguishing such numbers from the rationals. For example, the number $x = 0.12122122212222\ldots$ is irrational because it does not repeat by the rule for its construction. Again, whereas these two formulations are logically equivalent, they are not *mathematically* equivalent. In both of these cases, logical equivalence is certainly not the same as *identity*.

Mathematics consists in large part of elaborate tautologies. As the great mathematician Henri Poincaré pointed out, if mathematics consists merely of elaborate ways of saying **P** if and only if **Q**, then how are we to account for its power and effectiveness in describing the natural world? The answer to this seemingly perplexing question lies precisely in the observation that a tautology is ambiguous and that ambiguity confers depth. Remember that an ambiguity requires two frames of reference that differ from one another and are mediated by a single idea. In the statement "**P** if and only if **Q**," the **P** and **Q** are the two frames of reference. A rational number is a quotient of integers, or it is a repeating decimal. On the surface, these are very different ways of characterizing rational numbers, yet the symbol or statement "if and only if" says that there is a unitary idea that is being expressed in these two different ways. We have now moved up to a higher level, the real numbers, and at that level rationals can be identified in two ways—and we are free to use whichever one suits us best in a given situation.

A mathematical tautology is evidently not the same as a logical tautology. A mathematical tautology may well have non-trivial mathematical content and this content lies precisely in the ambiguous nature of a tautology, the fact that it provides two different frameworks in which to view comparable ideas or concepts. Now, these frameworks may be so close to one another that nothing mathematical is gained by the ambiguity. Many definitions fall into this category. An integer is even if and only if it is a multiple of 2. You might say that there is little content here because "even" *means* that it is a multiple of 2. Of course, if you merely listed the counting numbers: 1,2,3,4,5,6, ... and noted that every alternate one starting with the second was even, then making the observation that every even number is divisible by 2 would have a (very slight) mathematical content. Not all tautologies are equivalent from the mathematical point of view.

The previous paragraph makes the point that mathematics cannot be reduced to logic. If that were the case, then logical equivalence would be the same as mathematical equivalence. As we saw, the value of the technique of "contrapositive proof" is that it takes two statements that are logically identical yet mathematically distinct. Mathematical tautologies, on the other hand, may be extremely valuable, but their value lies precisely in their ambiguity.

Tautology or logical equality merely divides up the world of propositions into discrete sets whose elements may have a deeper structure—for example, the mathematical or the scientific, which can still be explored. Some tautologies are trivial or superficial, while others are very deep, and reframe a given idea or concept in new and surprising ways.

Ambiguity in Physics

So far, the ambiguities I have discussed are mathematical or logical. They are in themselves surprising, but only until you make a shift in your way of thinking. After breaking with a certain fixation on the formal logical structure of mathematics most people say it is quite obvious that the ambiguity of multiple perspectives is present in mathematics and accounts, in part at least, for its profundity. The mathematician Barry Mazur, for example, states in a recent paper that, "The heart and soul of mathematics consists of the fact that the 'same' object can be presented to us in different ways."[18] Non-trivial mathematical tautology, or multiple representations, is one aspect of the condition I am calling ambiguity.

One might argue that the examples we have enumerated up to this point only deal with the language and logical structure of scientific theory (in the sense that mathematics is the language of science) and not with the natural world that science is ostensibly describing. So I shall now turn to the natural world. Can we claim that the natural world is ambiguous? It is impossible to discuss the natural world without using some scientific concepts. Let's begin this discussion with quantum mechanics. According to the famed Copenhagen interpretation, still the dominant explanation of the subatomic world, an electron or any other subatomic particle has a curious existence, being both a particle and a wave. The physicist Nick Herbert highlights the paradoxical aspects of the situation:

> If we ignore observations for the moment, we might be tempted to say that an electron is all wave, since this is how it behaves when it's not looked at. However this description ignores the massive fact that every observation shows nothing but little particles—only their patterns are wavelike. If we say, on the other hand that between measurements the electron is really a particle, we can't explain the quantum facts.[19]

In certain experimental situations, the electron behaves like a localized individual object. In others, it has the wavelike properties of diffraction and interference. The electron (and other subatomic particles) appears to come with two descriptions—the particle and the wave. The situation appears to be paradoxical because these two descriptions are in conflict with one another. We want to say, "What is it really? Is it particle or is it wave?" Countless books and theories have been written to "explain" this paradox; countless arguments have raged over the years about the "true" nature of these subatomic entities and, by inference, the nature of physical reality itself. Our solution is simple—the electron is characterized by the total, ambiguous situation. It is the single entity that comes with two consistent but conflicting frames of reference. You don't have to eliminate the conflict because the conflict is an irreducible aspect of the subatomic particle.

The difficulties people have with quantum mechanics come not from the experimental results but from the interpretation of these results. It stems from our need to impose a framework of classical logic on a theory that has uncovered an aspect of reality that is more profound than classical logic can handle. The debate is not about the experimental results of quantum mechanics nor is it about the mathematical formalism that works extraordinarily well at predicting the outcome of experiment. It is about fitting this scientific theory into what could be called a classical framework and that just does not work.

The property of subatomic objects that I have been discussing is usually called complementarity, but this word is misleading. Normally, it refers to having two frames of reference that add up to one whole, but that is not what is going on here. Here, we have two descriptions, both of which give a fine description of reality but which appear to conflict with one another. Our problem is that we feel there is only one reality that is objective and clear, so we ask which description is correct. We try to resolve the situation either by imposing exclusive either/or duality or by complementarity, which is a kind of both/and. In fact, the situation is precisely what I have been calling an ambiguity. An ambiguous description succeeds in accounting for all of the features we have observed, *including* the seemingly paradoxical aspect. To sum up, according to quantum mechanics, the subatomic world, being the natural world at its most basic level, is characterized by precisely the sort of ambiguity that I have been describing.

$E = mc^2$

Einstein's equation $E = mc^2$ was used as the title of a book by David Bodanis.[20] In order to demonstrate the fundamental breakthrough that is contained in this equation, Bodanis discusses the ambiguity that the equation contains. He says that this equation (and all equations for that matter) is not to be understood as a simple balance confirming that "two items that you suspected were nearly equal really are the same." This equation has a metaphoric quality and encapsulates a fundamental insight into the nature of reality. Before the insight, the world was divided up into the disjoint domains of matter and energy. The division between matter and energy was absolutely fundamental to our cultural history and, as such, found its reflection in the very language we use. In a sense, it mirrors the more fundamental split between spirit and matter. Matter has mass and occupies space, but energy refers to force or activity or the potential for activity. Matter is expressed through nouns, but energy is closer in spirit to action, to verbs. They are the two consistent but conflicting frameworks that are called for in our definition of ambiguity. And these two domains are in conflict! If you look at this conflict as matter versus spirit, matter versus energy, objects versus process, or merely as nouns versus verbs, it is embedded deeply into any attempt to describe reality.

Now let's return to the equation $E = mc^2$ and ask what it reveals about this duality. It says energy *is* matter, where the "is" does not stand for identity but, in the way we have come to expect from a situation of ambiguity, that there is one reality that is matter when we think of it in one way and energy if we look at it in another. This is very similar to what I said about the fundamental theorem of calculus—that there is one calculus that is differential calculus from one point of view and integral calculus from the other. This implies that there is a way of translating from one domain to the other. In the case of Einstein's equation, the translation leads to the creation of atomic weapons or of atomic power. So the insight in this equation, that energy and matter form an ambiguous pair, does not stay at the theoretical level but has the deepest practical implications. An ambiguous reality is best described by a science that incorporates ambiguity at all levels.

The theories of physics are permeated with the sort of ambiguity that I've been describing. Electricity and magnetism are not distinct phenomena but aspects of one electromagnetic field. This insight by Michael Faraday and James Clerk Maxwell has justly been called the highlight of nineteenth-century scientific thought. Among other triumphs, it illuminates the nature of light itself as an electromagnetic phenomenon.

Ambiguity in Biology

For the most part, biology is a "classical" science.[21] By this, I mean that the dominant view in biology does not admit that situations of ambiguity or complementarity are legitimate. Nevertheless, some voices take another view. This tradition goes back to physicist Niels Bohr. As Freeman Dyson puts it,

> Bohr in 1932 proposed to extend the idea of complementarity to biology, suggesting that the description of a living creature as an organism and the description of it as a collection of molecules are also complementary. In this context, complementarity would mean that any attempt to observe and localize every molecule in a living creature would result in the death of the organism. The holistic

view of a creature as a living organism and the reductionist view of it as a collection of molecules would be both correct but mutually exclusive. Bohr believed strongly in this application of complementarity to the understanding of life.[22]

In the same article, Dyson refers to the work of Carl Woese, the biologist I introduced in chapter 2,

> Woese's new biology is based on the idea that a living creature is a dynamic pattern of organization in the stream of chemical materials and energy that passes through it. Patterns of organization are constantly forming and reforming themselves. If we try to observe and localize every molecule as it passes through an organism, we are likely to destroy the patterns that constitute metabolic life. In Woese's picture of life, complementarity plays a central role, just as Bohr said it should.

The attempt to develop a view of biology that is non-reductionist and consistent with its evolutionary and dynamic nature will inevitably necessitate the introduction of some consideration of ambiguity. The essence of complementarity, as Bohr uses the term, is, I maintain, its ambiguity.

CONCLUSION

A number of quick inferences can already be drawn even from this preliminary discussion of ambiguity. The first of these is that, contrary to what most people think, the job of mathematics and science is not to banish ambiguity which, in certain situations, adds to the richness of a scientific situation. The more ways you have to look at a given situation, the better you understand it.

The role of ambiguity in science is akin to the role of metaphor in poetry. Metaphor is where the richness of poetry resides, and in science it is what accounts for the depth of scientific ideas and concepts. Mathematicians love to talk about "deep" ideas in math as opposed to "trivial" ideas. What makes something deep? What makes it trivial? These are the kinds of questions that naturally arise from the discussion so far. In a similar way, you can differentiate between the words complex and complicated. Complexity, for me, contains a conceptual depth, whereas something can be superficial and yet be quite complicated. A comment made by Gregory Chaitin[23] is relevant here: "We have computers now, so we don't have to have people imitating machines." Human creativity involves ideas, ambiguity, paradox, depth, and complexity. Machines live in a formal world that is very complicated but lacks depth. The depth is to be found in the human beings who create the computer world. Depth lies in the ambiguity of an idea—the fact that ambiguity is generative, that it is multi-dimensional and comes with multiple viewpoints.

A "representation" can be thought of a particular point of view. The philosopher Emily R. Grosholz emphasizes the importance and ubiquity of this kind of ambiguity in her excellent book, *Representation and Productive Ambiguity in Mathematics and the Sciences* (Oxford University Press, 2007). Grosholz discusses the importance of such multiple representations for problem solving in science. She undertakes a historical analysis of the "productive ambiguity" of seminal texts by scientists like Galileo, Descartes, Newton, and Leibniz, in scientific domains that vary from molecular biology and genetics to logic and topology. Her work has the potential to radically change the way we normally think about the nature of science and mathematics.

What is the normal response to ambiguity? This question brings to mind an article that recently appeared in *The New York Times*[24] entitled "Go Ahead, Rationalize. Monkeys Do It, Too." It's about some research at Yale that found evidence of cognitive dissonance in monkeys and in four-year-old children. The article contained this interesting sentence, "In general, people deal with cognitive dissonance—the clashing of conflicting thoughts—by eliminating one of the thoughts."

Evidently, monkeys also react to cognitive dissonance in this way. After all, dissonance is the essence of ambiguity. The conflict that I spoke about generates a tension that makes people feel uncomfortable. They want stasis, equilibrium, and tranquility. So they suppress one aspect of the ambiguity. This is what the Greeks did with zero and the root of two. This is what logic does for us. It enforces consistency by attempting to banish ambiguity but you just can't do it permanently. Why? Because ambiguity, not logical consistency, is the way things are. The true nature of mathematics and science is consistent with the nature of the real world.

6

⤳

Self-Reference:
The Human Element in Science

DEEPER INTO AMBIGUITY

Science is divided from life by a built-in gap that is not sufficiently appreciated. Life is lived from the inside out as our subjectivity reaches out, as it were, to the objective world. Science goes in the opposite direction. It brings the outside world into our subjective comprehension. This chapter will bring this gap into sharp relief by introducing the ambiguous situation of the scientist whose presence hides unacknowledged behind most discussions of science. The blind spot and the resulting complexity of my description of science is intimately connected to the need to include the self-consciousness of the observer in a description of science. It is the ambiguity that is implicit in the human condition itself, the ambiguity between objectivity and subjectivity, which is the subject of this chapter.

The previous chapter discussed the way in which ambiguity appears in science and mathematics, in its structure, operations, concepts, and theorems. Even at this stage the existence of ambiguity forces us to change the way we think about science. Instead of a science that is characterized by certainty, solidity, and precision—the logical but rigid world of the formal theory—the discussion of ambiguity demonstrates that there is an unexpected flexibility to science and mathematics.

For most of the ambiguities of the last chapter, a "higher" point of view was cited, from which the ambiguity was resolved. As a result, the situation is essentially static. Even if use of the word ambiguity in the context of a discussion of science may have initially seemed strange, this form of ambiguity is relatively straightforward. We tend to put ourselves in the position of the expert—the one who "understands"—for whom the ambiguity has been resolved.

Initially, ambiguity is most surprising when it is used in reference to the logical structure of mathematics. Most of us are most comfortable when thinking about the objective content of a subject, because this enables us to hold the matter at arm's length. Unfortunately, if we are to gain some insight into ambiguity, and thus into the novel approach to mathematics and science that I am proposing, holding ambiguity at arm's length will not work. In this chapter, it is precisely our cherished "objectivity" that will be put under the microscope.

In a way that is analogous to my discussion of subjectivity in chapter 2, the word "objective" is also quite subtle and ambiguous. It can mean "not subject to personal feelings or opinions,"[1] and in this sense, science and mathematics are objective. It can also mean "not dependent on the mind for existence."[2] I shall usually be referring to this latter meaning of the word when I use the term "objective," and in this sense the objectivity of science can certainly be questioned.

We can begin to see that ambiguity does not just reside in the objective content of science when we recall the discussion of the differing perspectives of the student and the teacher. We saw that the same situation was not ambiguous for the teacher, yet it was for the student. In fact, a mathematical situation may be ambiguous for one student and not for another. It's not just that one student may "get it" and the other may not. It is also that one student may just not perceive that there is a problem at all! "Yes a product is both process and object. So what!" It may well be the more mathematically sensitive students who will be bothered by the ambiguity. They will see the conflict in the ambiguity as something they must resolve—that just cannot be left alone. The less sensitive student may not see there is a conflict or even that there is anything wrong with having two conflicting views.

Learning does not happen unless there is an acknowledgement, implicitly or explicitly, that there is a problem, question, or deficiency. Learning is not merely the accumulation of facts. It consists of an insight into the relationship between facts. It arises with the emergence of an idea that answers the question, "What is going on here?" The sensitive student, the one who sees a problem in the situation, who perceives the ambiguity, is the one who has the opportunity to learn the most from the given situation. You must "not understand" something before you can understand it. "Not understanding" is not always a pleasant condition to be in. It makes most people uncomfortable. It's much more pleasant to be in control, the master of the situation. It takes a kind of courage to acknowledge that you don't understand, but without this acknowledgement it is impossible to learn or create anything of value. Yet, "not understanding" can be exciting and challenging. Learning something new is fun!

Ambiguous learning situations contain a conflict between theories or points of view that cover some of the same ground. Learning involves the creation of the single idea that integrates the different viewpoints and resolves the conflict. This integrated, flexible, and inevitably higher-level point of view brings with it a feeling of ease and a new sense of command of the situation. Learning is dynamic, it has a before and an after, and this is true for any situation of ambiguity.

These considerations imply that ambiguity is not an objective element of the situation. *It must be perceived!* This brings me to the consideration of another kind of ambiguity that is not usually talked about, namely the one in which the objective content of mathematics or science is confronted by the subject who creates or learns it. Objectivity and subjectivity are themselves conflicting frameworks that inevitably come into play in any discussion of math and science that aspires to give a complete picture of what is going on.

THE PROBLEMATIC NATURE
OF THE PHILOSOPHY OF SCIENCE

This new kind of ambiguity is at play when I write about science. I write both as insider and as outsider—both as a scientist and as a meta-scientist or philosopher of science. As a philosopher, I am an observer. I am standing outside of science and asking what it means, how it is created, whether or not it is true, and so on. I view the subject as a whole but I view it from the outside. Many scientists never think about science as a whole. Nevertheless, acting as observers is part of what they do when working in science.

It is also the way most mathematics, for example, is taught, at least for those who operate in a formalist mode. It is as though the teacher is showing the student a picture of some beautiful scene. Both students and teacher are observing the scene. The teacher sees the scene in greater detail than the student and this leads many students to conclude that what is being asked of them is merely to become familiar with the picture in as much detail as possible. As long as the mathematics stays on the outside, as long as the student remains an observer, there is no learning going on.

Another mode of consciousness is at play in mathematics and science that is central to the description of mathematics that has been developed by people like Imre Lakatos[3] and Reuben Hersh,[4] both of whom emphasize the social and historical dimensions of mathematics. It is called the "participant mode." As a scientist, I am also a participant. I am a member of the scientific community and immersed in its culture and traditions. When we emphasize that science is a human activity, we are stressing the participant aspect of science. This participant point of view is so strong there is reluctance on the part of most mathematicians and scientists to stand back and take a good look at what they are doing.

The study of texts or theories that include the participant in the study is called "hermeneutics." Usually hermeneutics involves the attempt to understand things from someone else's point of view and to appreciate the historical, cultural, and social forces that may have influenced their outlook. This is not the way we usually look at mathematics and science but it does provide a new and interesting perspective, especially in comparison to the usual way of thinking about science, which is from the point of view of the observer and, for this reason, embodying an objective and absolute truth. In a sense, a humanistic philosophy of science would be a hermeneutics of science.

One of the principal characteristics of any culture is that it has a set of rules that define what is meaningful. There also exists a meta-rule (this is essential) that the cultural assumptions are real and true—in fact, the culture *defines* what is real (within that culture). We are participants in a culture to the extent that we accept that culture as real. This is as true for scientific culture as for any other. I write and people read from within the culture of science. Every time I bring up something that varies from the dominant cultural assumptions, I must do so very slowly and carefully. This is the reason I introduced ambiguity the way I did. A scientist works as a participant in scientific culture. The two modes of consciousness, observer and participant, are very much present in every part of science. One of the myths of science is that the participant mode can be dispensed with. In mathematics, we also have the idea that the subjective participant mode is irrelevant. Both of the dominant attitudes toward mathematics—that the truths of mathematics exist in some ideal realm (Platonism) or that they must exist within a logical axiomatic system (Formalism)—are completely objectivist, with no participant—no mathematician—in sight.

These observations may help explain why the philosophy of science is intrinsically difficult and problematic. Most academic discourse is grounded in the point of view of the observer. But a successful philosophy of science must be written from both the inside and outside of the subject. Many philosophers, even those who write about science, do not know science from the inside, and similarly scientists usually know little philosophy. Even if this were not the case, there would still be a problem—*it is intrinsically difficult to simultaneously adopt the position of observer and participant.*

Albert Low[5] uses the performance of a stage actor to highlight the inherent conflict between observational and participatory modes of consciousness:

> Assume that you are an actor on the stage looking out at a member of the audience. Now let us assume that you are a member of the audience looking at the actor on the stage....
>
> To be an actor is to be involved, to take sides, to care about the outcome. It means that one must put out energy, have goals to be able to succeed or fail. Yet we are not only actors, we are also observers. As observers we can be cool, detached, impartial, uninvolved, relaxed. To be involved is to see the world as though from the inside; to be an observer is to see the world as though from the outside.

Teachers often teach from the observer point of view precisely because in it one can be "detached, impartial, relaxed"—in short, in control. Unfortunately, the student rarely sees their teacher as participant. For a number of years, a colleague and I taught a course called "Great Ideas in Mathematics" in which we introduced students from a wide variety of disciplines to what mathematics feels like from the inside. We based the course on the historical approach developed by William Dunham in his wonderful book, *Journey through Genius: The Great Theorems of Mathematics.*[6] One of the features of the course was that both professors were present at all times. We would take turns presenting the material and moderating the discussions. The person who was not at the front of the class felt free to interject his own questions or comments and these often led to energetic debates between us or between either one of us and the students. The students often commented on how much they enjoyed the dialogue between the two instructors. From their responses we learned that they had rarely if ever witnessed people discussing mathematics, and especially not discussing the historical, cultural, and philosophical significance of mathematical results. Through this dialogue they learned about what participating in mathematics might be like and they found this to be an exciting eye-opener.

Students rarely see their teachers in a situation where the teacher is not in complete control, where things do not make sense or the answer will not come out right. Even if the teacher has her own idiosyncratic way of looking at the situation or her own take on some concept or theorem, she may not share this with her students. The reason for this is that these things seem too subjective and the teacher feels that her role involves conveying the *objective* mathematical content—the facts and nothing but the facts. The participant mode of being is often hidden from the student (this may not be as true for advanced graduate seminars but it is certainly true of most research papers), yet without entering into what I am calling the participant mode there can be no learning and no mathematical research. Creativity in its largest sense is a function of both modes of consciousness.

The suppression of the subjective participant way of thinking is not done out of a perverse desire to cause problems—there is a very good reason why the subjective is kept out of sight. In his writing, Low stresses that the viewpoints I have been calling the participant and the observer form an ambiguity that has its roots in human nature itself. Human beings have two distinct ways of being in the world—that is, two different modes of consciousness: those of actor and observer, precisely. These two modes don't just sit benignly side by side. They are not complementary. They are ambiguous! Calling them ambiguous brings out the conflict between them. If the stage actor begins to observe her own performance, she may well become self-conscious, which may affect her performance or even bring on stage fright. There is something intrinsically disturbing about being in these two modes of consciousness simultaneously. Standing outside of yourself and regarding yourself as an object, brings about a situation of self-reference in which one is split down the middle, an existential state that can be uncomfortable to say the least. There is a tendency for such situations to get out of hand. Self-referential situations often lead to logical paradoxes that are uncomfortable especially for people like mathematicians who are great believers in consistency and are masters of logical reasoning. To be totally a participant is to be active, vibrant, and alive. To be an observer is to be cool, detached, comfortable, and in control. To alternate between the two is a delicate balancing act that can threaten to get out of control. To be both simultaneously is threatening and frightening—in short, impossible.

This situation is the heart of the problem with the philosophy of science. If one writes from the inside, so to speak, one is not a philosopher; write from the outside, and one is not a scientist. Thus, the scientist may be reluctant to write about science for fear that it will endanger their status as participant. Perhaps it is not that exactly. Perhaps it is that even thinking about science as an observer is threatening to one's status as a participant. It may be that "the unexamined life is not worth living," but the very act of looking at something from the outside causes a loss of innocence—in other words, a person will never again assume the same kind of unquestioning attitude toward science that they had as a full-fledged and unquestioning participant.[7] In science, this situation is ironic to say the least. The attitude of science toward reality is precisely that the participant mode of consciousness is less real—less true—than the observer mode. This is held to be true of everything *except for science itself.* When it comes to science, the role of the observer is problematic.

Most discussions of science are framed using the point of view of the observer. While a few are biased toward the participant side, none recognizes the ambiguity inherent in the situation. Recognizing this unavoidable ambiguity that arises whenever we do science leads us into an extreme and perplexing complexity. This is where I am now headed. The difficulty is substantial but the rewards are a viable approach to the actual situation.

What are the implications of the observer/participant split for the discussion of science? Consider the role of proof in mathematics (the same considerations apply to many situations in science in which experimental results are formalized for publication). You are working on a problem and get some idea, an intuition that something systematic is going on. The next step is to try and write it up. This amounts to reducing the mathematical situation to one that is objective. To do so you move into your "observer mode"—you put yourself in the place of someone whom you want to convince that an idea is valid. All kinds of difficulties may arise at this stage that may lead to reworking parts of the argument or even reconceptualizing the original situation. It may lead to entirely new ideas that are then formalized in an iterative process that ends when a result appears that is satisfactory to the researcher. What is happening here is an alternation between the observer and participant modes of working. Every step in the learning or creation of science involves an interaction between these two modes.

When we say something is objective, we may think this means it is independent of the observer; that it would be true even if there were no observer around. This is the reason for calling scientific results "laws" rather than "patterns" or "regularities." Yet in practice there is no situation that does not come with a point of view, and an observer has a point of view. "Independence of the observer" really means independence from what I called the participant mode of consciousness; it remains totally dependent on the observer mode. Objectivity is one way of looking at the world. Similarly something is subjective if it arises from the participant mode. In the same way that the word "objective" can be used in two different ways, similarly "subjective" has the same two meanings: either "dependent on personal feelings" or "dependent on mind." I am not interested in the former meaning, even though one can certainly be a participant in this sense, as when one gets emotionally involved in a discussion. Paradoxically, the subjectivity of the mind is objective in the sense of it being independent of personal opinion and prejudice.

The positions of observer and participant are not symmetrical. You can observe (become "aware of") your state as participant, but you cannot participate in your observation. Notice that (indulging in an irresistible mathematical reaction to this situation) it is possible to iterate this situation:[8]

- You can be a participant.
- You can observe (be aware of) yourself as participant.
- You can observe your observation of yourself (be aware of being aware) as participant.

And so on....

Participation stands alone, but observation can be iterated. This would seem to be a fundamental difference between subjectivity and objectivity; the former is unitary and foundational, whereas the latter is intrinsically bifurcated and therefore capable of a kind of infinite refinement. Nevertheless, a conflict exists between them and one might argue that the iterated sequence mentioned earlier is a direct consequence of attempting to reconcile the conflict generated by being a unitary being with two conflicting modes. (Unity, as the basic condition that generates all this, is discussed in detail in chapter 9.)

A New Classification of Science

This spiral of higher order and more tenuous observations points to a new way to look at science, a different kind of classification of scientific activities. This will be based on the extent to which both observational and participatory modes are acknowledged. Often, the modern view of science is that it is a series of increasingly refined approximations of reality. Here, we have a different kind of approximation springing directly from the inevitably bifurcated view of the scientist.

The first and most basic level I shall call "classical science." It is "normal" science—how most of us think about science—totally observational and objective, characterized by the denial of subjectivity of any kind. There is no hint here of the participatory mode of consciousness. In a subject like Newtonian mechanics, we stand back and look at the universe from the viewpoint of a hypothetical observer whose vantage point is outside of the universe we are describing. Even discussing a subjective state like consciousness from a scientific point of view is usually done by reducing things to objective properties of the brain.

The next level is an acknowledgment that the experimenter is a human being and therefore is implicated in the experimental arena. Consider, for example, a social science like anthropology. The anthropologist who is studying some so-called primitive tribe thinks of herself as an observer but is aware that her acts of observation have an effect on the situation. In this way, the anthropologist is a participant. The loss of one's status as pure observer is sometimes referred to by the pejorative expression "going native," which refers to a fall from the ideal of total objectivity. Of course, if the researcher does not enter fully into the situation she is studying, she may never succeed in understanding what is going on. Faced with this dilemma, she still attempts to be a pure observer—to establish the state of affairs independent of her presence. But this may not be possible. All she can do is ignore the effect she has on the situation or make up some ad hoc method of discounting those effects.

This is a problem for all of the social sciences—in fact, for all of the sciences that study human beings in one way or another. Economics may regard itself as a classical science through the use of sophisticated mathematical models, but it cannot avoid such self-referential complexities. Take the stock market. The essence of the problem with studying the stock market is that the people who are studying the market are simultaneously participating in it. One might say that the effect of any single individual is negligible but not so the effect of vast numbers of observer/participants. This is the reason that the market always overreacts, and is always consumed by the contagious emotions of fear and greed. Everyone is trying to get some personal advantage by out-guessing everyone else who is trying to get the same personal advantage.

So we can see that all of the human sciences are not classical because the participant status of the scientist cannot be reasonably ignored. Even so, there is always a tendency for the scientist to revert to the status of pure observer. Statistical techniques are often the means through which this is done. Even if the effect of the scientist is acknowledged, one keeps trying to give an "objective" accounting of the situation. One operates "as though" such a condition of absolute objectivity exists. It is exactly this ultimate state that is put in doubt by this analysis.

One might think that we have isolated the very factor that distinguishes the social sciences from the natural and biological sciences. Yet the same considerations are not absent from the natural sciences. I have already discussed (in chapters 2 and 5) the ambiguous notion of complementarity in quantum mechanics, but it may be useful to add a few more comments to that discussion.

In a sense, quantum mechanics is a problematic theory precisely because it raises the question of the status of the scientist. Is she an observer? Is she a participant? Does she alternate between being observer and participant or is she both simultaneously? Is there any place in all of this for the scientist as pure observer as classical science demands?

Quantum mechanics can be considered a classical theory. It makes predictions that can be verified to a high degree of numerical accuracy. It can be applied to concrete situations and produces concrete results. It has a mathematical formalism. At this level, it is an observational theory. The scientist stands outside of the experiment and describes what is going on; meaning is not the issue. Consider the following remarks by the Nobel Prize–winning physicist Richard Feynman in his book about quantum electrodynamics:[9] "It is my task to convince you [the reader] *not* to turn away because you don't understand it .... that is because *I* don't understand it. Nobody understands it .... while I am describing *how* Nature works, you won't understand *why* Nature works that way. But you see nobody understands that." "How Nature works" refers to quantum mechanics as a classical theory.

However, quantum mechanics contains a well-known twist, an additional complexity that was introduced in my discussion of the wave/particle ambiguity in chapter 4.

Nick Herbert said about the electron:

> It looks like a particle whenever we look. In between, it acts like a wave. Because [a] measured electron is radically different from [an] unmeasured electron, it appears that we cannot describe [the electron] as it is without referring to the act of observation.[10]

The problem here is what is called "the collapse of the wave function." This is the claim that the state of a subatomic particle such as an electron consists of a probability wave consisting of all possible states until an act of observation triggers the "collapse" of this wave and the emergence of one definite state. This is not a question of lack of knowledge since one cannot, *in principle*, determine which state the particle will assume. Before the act of observation, it is not even reasonable to make the statement that the particle is in a specific state. It is the act of observation that triggers this "collapse." The scientist is therefore not outside the experiment. She is on the inside affecting the outcome by her act of observation. So in this sense you could say that quantum mechanics is a participatory theory. Questions of meaning, the "why" that Feynman refers to, can best be addressed at this level.

If the ambiguous situation I have just described is brought to the fore most strongly in the social sciences, and it also exists in particle physics, there nevertheless remains an element of it in all of science. The reason for this is because the participant/observer split is fundamental to the human condition itself, and so inevitably the scientist carries this split into her scientific work. The very claim that science is objectively true is itself a response to this ambiguity. Therefore, there can be no philosophy of science that does not take this condition into account. All scientific experimental situations have this element of observer/participant ambiguity. Acknowledging one's role as participant and observer precipitates an infinite cascade. Each of lines 1, 2, 3, … (on page 99) refers to a mode of consciousness of the scientist. Normally, the scientist works from (what he imagines to be) a single mode of consciousness (observer). It gives a unique perspective (point of view) from which to understand and communicate science. To admit his role as participant threatens to destabilize the entire scientific process. Yet this is the reality of the situation.

In his famous book, *The Structure of Scientific Revolutions*, which I referred to in chapter 4, Thomas Kuhn compared "normal" with "revolutionary" science.[11] My distinction between classical and self-referential science is certainly not the same as his, but it is interesting to compare them. Kuhn's "normal" science takes place within a well-established paradigm. In such a situation, the status of the observer is not usually a problem. So most normal science, especially the examples discussed by Kuhn, would be classical. Kuhn's revolutionary science occurs when a paradigm is found to be deficient and there arises a competition between alternative paradigms. At such a time, the participant status of the scientist is certainly relevant. What Kuhn calls the "incommensurability" between paradigms arises precisely because it is not possible to participate simultaneously in two distinct paradigms. After all, the paradigm is the framework within which you see the world. In this regard, Kuhn[12] quotes physicist Max Planck, "a new scientific truth does not triumph by convincing its opponents and making them see the light, but rather because its opponents eventually die, and a new generation grows up that is familiar with it."

One of the things that was so original about Kuhn's approach to science was that he took issue with the naïve idea of continual scientific progress toward an objective and unique description of reality. The famous introduction of "paradigm" into a discussion of science pushes the human element to the fore.

When it comes to mathematics, the dual roles of participant and observer may, at first glance, not be as clear and straightforward as they are in the situation of a scientific experiment. It is more like the situation of a theoretical scientist than that of an experimentalist. Nevertheless, the same observations are relevant. The student initially observes the mathematics as their teacher or textbook presents it. They may only move into the participant mode when they are back home. This may occur when they are trying to understand the material that has been covered in class, or when attempting to solve problems. The strange thing is that because the formalist approach to mathematics highlights the observer mode, the student often feels that writing things in a logical sequential order is *the* way to understand math or to do problems. Of course, "understanding" means precisely the opposite: It means developing a subjective feeling for the material. It is this subjective feeling that is subsequently given an objective and conventional form. The same discussion applies to the research situation that involves a continual interaction between the objective and the subjective, between intuition and logic, between participant and observer.

CONCLUSION

The view that I have called "observation" is the basis for objectivity, the participant view for subjectivity. The claim that science is objective is really the claim that one of these modes of consciousness is more valid than the other. Remember that I mentioned that the word "objective" has two distinct meanings: independent of personal opinions or impartial; and independent of mind. Science confuses these two meanings. It identifies the valid goal of impartiality with the invalid goal of being independent of mind. In so doing, it tends to disqualify all consideration of the participant, replacing participant by observer, not realizing that the participant point of view is also an objective phenomenon (if you take the "impartial" sense of both of these terms.) The position of the ideal scientist is that of an impartial and omniscient observer, a kind of God, standing outside of the phenomenal universe and observing its patterns and regularities.

One of the aims of this book is to unify science and meta-science. Most mathematicians and scientists are afflicted by a sense of vertigo when they try to look at things from these two viewpoints simultaneously. That is why there is so much impatience amongst scientists toward the philosophy of science. I believe that the philosophy of science can only legitimately take place within the ambiguous context that encompasses observer and participant. The trite comment that the scientist knows no philosophy and the philosopher knows no science should be taken to mean that the scientist lives primarily in one context, while the philosopher lives in the other. Both in their own way have abolished the ambiguity inherent in the situation, and so both science and the philosophy of science go their own way independently of one other, in a kind of mutual ignorance and contempt. Actually, both sides need each other. Clearly, a philosophy of math without math is sterile. But mathematics without meaning is doomed to its own form of sterility. This was the danger of formalism, and even today this approach to the learning and teaching of mathematics threatens to destroy the subject.[13] The sad thing about the philosophy of science is the idea that it is the role of math and science to banish ambiguity, when precisely the opposite is true: Without an appreciation for the inevitable ambiguity of the situation, we shall never be able to understand the deeper significance of science and math.

One of the most basic aspects of the human condition consists in the fact that our consciousness has the ability to operate in the two distinct modes of actor and observer. It is the incompatibility between these two states that leads to various attempts at reconciliation, an attempt to re-create a unified point of view out of this primal bifurcation. Science and mathematics are two of the most powerful methods through which reconciliation is attempted. As is usual in any ambiguous situation, there are two ways to proceed. The first solves the problem by repressing the participant mode; asserting that everything is external, objective, and essentially unchanging. The second involves allowing oneself to become aware of the ambiguity of the situation and using that ambiguous situation as a springboard to generate new science and mathematics. The second option consists of turning ambiguity into creativity. The proper role of the philosophies of mathematics and science does *not* consist in replacing the complete human condition by some reduced and partial version in order to allow everyone to relax and take it easy. Any legitimate philosophy of science should be concerned with revealing the actual situation in all its ambiguous complexity and dynamic power. If philosophy is concerned with a study of reality, the philosophy of science and mathematics should give close consideration to how the human condition reveals itself within science and math. Science, on the other hand, investigates the same ground as philosophy: human nature and the nature of the natural world.

7

∞

The Mystery of Number

It is paradoxical that while mathematics has the reputation
of being one subject that brooks no contradictions, in reality
it has a long history of successfully living with contradictions.
This is best seen in the extensions of the notion of number
that have been made over a period of 2,500 years ... each ex-
tension, in its way, overcame a contradictory set of demands.
—Philip Davis[1]

The last chapter was about limits to analytic intelligence. These limits are not just global, they are also local. They come up in almost
every situation in which science is discussed; they are implicit in every
basic scientific concept. Every scientific theory is supported by a web of
primordial ideas; notions such as time, space, number, energy, and matter.
Without such ideas, we could not get started with our scientific investigations since these building blocks help us develop a language in which
to express our theories, laws, and equations. In a very real sense, such
basic ideas generate the scientific world. Ideas such as time and space
may seem obvious, but do we really understand these things? And what
would understanding them actually mean? The question addressed in
this chapter is whether we can find a way of talking about scientific concepts that is consistent with the existence of the point of view I have
been calling the science of wonder, whose uncertainty comes from the
existence of limits to reason. To do this, I now introduce the notion of a
proto-concept and use it to discuss "number."

PROTO-CONCEPTS

Understanding something like time, space, or number is an interesting and subtle affair. The basic building blocks of science are not and cannot be defined (definitively). For this reason, I call them proto-concepts. They are dynamic generators of families of concepts. If you think of a proto-concept as a crystal, the defined concept is a face of the crystal. In this section, I am interested in differentiating between the crystal and its faces.

Let us take up "number" as an example of a proto-concept. What does the term "number" refer to? The Pythagoreans believed that "the world is number," and they were certainly on to something important—a point of view that is possibly even more significant today. They certainly did *not* mean that the world was identical to some particular number or even to some set of numbers. Perhaps they meant that there is no aspect of the natural world from which number is absent. To the Pythagoreans (and many others), number represented the principle by which the world is ordered. The great mystery about the natural world is that it is intelligible at all and not completely chaotic. The existence of order is the foundation on which all structure, all language, science, and even these observations, are built. The most elementary way in which data are ordered is through the observation that there are seven objects, for example, instead of just saying there are "some" or "many." The most obvious and powerful way in which situations are structured is through the use of numbers.

"Number" is basic to our civilization. Mathematics has been called the "science of number," and even though such a characterization has an old-fashioned ring to it, it is still present in other ways of talking about mathematics, such as the "science of pattern." The patterns we study in modern mathematics are, for the most part, numerical in nature. Whether the subject is geometry or algebra, probability or combinatorics, the theoretical edifice is usually built with reference to some numerical domain, which is most often the real or complex numbers. So mathematics itself, as well as all of the pure and applied sciences that use mathematics (and that includes almost all of them), is based on the foundation of "number."

And what *is* a number? This is not an easy question. The concept of number is not fixed and absolute. It has evolved over the centuries, and as mathematician Phillip J. Davis observed in the quotation that opened this chapter, each advance involved overcoming a contradictory set of demands. Number evokes a definite experience of interacting with the natural world; it points to some definite aspect of reality. It structures both the natural world and our way of thinking about the natural world. I want to find a way of seeing and talking about "number" that respects its developmental nature as opposed to a description that is fixed and absolute. That is not to say that it is impossible to define particular kinds of numbers based on, say, certain axioms. Nevertheless, it is number as a proto-concept that produces these different axioms, these different formulations.

Quantity

Instead of rushing into a definition of number, let's ask instead if there is something that is more elementary than number—something that might have given birth to the idea of number. Aristotle claimed that number arises out of the notion of quantity. Is "quantity" identical to "number"? It would seem that quantity is a precursor to number since it is difficult to imagine the notion of number without that of quantity. Maybe number is a way of getting a handle on the experience of quantity. But then what is quantity?

When you look around at the natural world one of the most basic things that strikes you is quantity. Quantity is the amount of something. It is measurable or potentially so. Measurement is the most basic thing you do in science; if something cannot be measured, I'd be tempted to say it is not science. You can scarcely imagine a scientific experiment that does not measure some quantity or other. And how is it measured? It must be measured by using a yardstick of some kind. And what is a yardstick? It is a standard model of the quantity being measured—length or weight, whatever. This yardstick is a way of introducing number into the situation. Notice that though this trio of quantity, measurement, and number is the foundation on which much of science is built, it is difficult to break this triple up into separate and distinct concepts. All are different aspects of what is essentially one situation.

Quantity is the latent mathematical potential that is implicit in a given situation, while measurement is the activity through which that mathematical potential is made explicit. One might then say that number is the actual mathematical content that has now been made explicit. However, if we wished to focus more on number as a proto-concept, we might think of it as the entire process through which mathematics manages to order the world. In this view, quantity, measurement, and "number as object" would form aspects of the proto-concept of number.

Quantification is potentially present in most situations. If it were not for quantity, then every object would need to be considered on its own—to be incomparable. In one sense, everything *is* unique and incomparable, but if we stop at this point, then there is no language and certainly no science. Before you can quantify, you must be able to compare. In order to compare, you must have some *thing* to compare.

One of the most basic functions of perception is the ability to differentiate figure from ground. In those ambiguous pictures associated with gestalt psychology, the "old woman and the young lady"[2] or "two faces and a vase," what you see depends on what part of the picture you take as foreground and which as background. When I look out of my study window at a winter's scene, the part of the scene I am focusing on becomes the foreground; everything else sinks into the unattended background. The notion of figure and ground emerges out of the act of perception. When we look at the winter scene—at the trees, the bushes, the rocks, and the sky—we see it as broken up into objects—we don't see a chaotic undifferentiated whole. The elements of the scene are *potentially* objects (when we attend to them) and an undifferentiated background (when we do not). The experience that a given visual situation can be broken down into units is a part of what goes into the idea of quantity. Notice that here I have connected "quantity" with acts of perception, which brings up the interesting question of whether quantity is objective or subjective. Even at this early stage one can see this is not such an easy question.

From the isolation of individual objects, we move to the "naming" or classification of objects. Our mind seems to function through the creation of abstract categories, for what is a "tree" after all but a set whose members are "trees and only trees." Thus, language involves classification through abstraction. In order for there to be two trees, for example, there must first of all be something that these two objects have in common, something we could call "tree-ness." No two trees are completely identical. There may well be some aspect of their DNA that characterizes "tree-ness," but for the most part when we talk about a "tree" we mean that all trees have in common their "tree-ness."

The situation is a little different when we talk about something like a "hydrogen atom." From the scientific point of view, all hydrogen atoms *are* identical and interchangeable. One could ask whether they really are identical. Every snowflake is reputed to be unique, but we assume that atoms, the basic constituents of matter, come in a finite number of varieties, and that each variety, like hydrogen, consists of identical particles. It is interesting that much of science involves the isolation of categories of objects—varieties of subatomic particles, atoms, or molecules—that are assumed to be absolutely identical. The isolation of categories of objects is a prerequisite to the concept of quantity.

The natural numbers are elementary examples of such a classification process. Collections of identical objects have quantity, and so we can say there are two objects, or three or four. Thus, after we have isolated the idea of "tree-ness," it becomes possible to talk about "two trees" or "three trees." We may start with the idea of the natural (counting) numbers and trace it back to the more elementary idea of quantity.

Quantity as Magnitude and Multitude

Quantity is not only a precursor to the counting numbers; it is also the factor that allows one to make comparisons between objects. It is related to the ideas "more and less," and the variants, "bigger and smaller," "longer and shorter," and so on. Is one line longer or shorter than another? Is one field bigger or smaller than another from the point of view of the amount of grain it can produce? Bigger, smaller, more, and less, are all properties of measurement, and measurement implies there is something to measure. This something we could also call quantity.

It is interesting that quantity has the two aspects, *multitude* and *magnitude*. Multitude is the factor behind the *discrete*, and discrete refers to a collection of objects that are "individually separate and distinct,"[3] a set of individuals. I shall discuss this in more detail in chapter 9, and it is exemplified by the counting numbers. However, when I discussed variants of "more or less," I was talking about "quantity as magnitude," which leads to the continuous, as in the *area* of a triangle or the *length* of a line segment. Quantity comes in these two varieties, and it is naturally ambiguous. It comes with two built-in frames of reference and there is a natural tension between these frames of reference. One leads to geometry and the other to arithmetic and algebra. One initially thinks of number as arising from counting—that is, from multitude, from the discrete and the algebraic. Yet because quantity is also magnitude, number also has the dimension of the geometric and the continuous. Quantity has these dual frameworks, magnitude and multitude, but is not completely defined by either framework since quantity is one thing. Quantity is "amount," the numerical potential of a given situation.

Multitude is not magnitude. We could give up here and, in the interests of clarity, leave magnitude and multitude as different, almost mutually exclusive concepts. If we were to do this, then we would never understand quantity nor would we have developed "number" in the way we have. The whole history of mathematics and science would be radically altered—there would probably have been no science as we understand it today. To grasp the concept of quantity and, in so doing, approach the profundity of "number," we must ask what quantity as magnitude and quantity as multitude have in common. To be circular about this, quantity is what they have in common. Quantity, you could say, is the field of meaning that emerges out of the ambiguity or polarity we have identified as magnitude and multitude.

We might begin with number as multitude. This leads to the "counting numbers"—one, two, three, and so on, and from there, directly to the development of the rational numbers or fractions. However, if you are a Greek geometer like Euclid or Pythagoras, counting is not the primary source of numbers. Numbers are tied to geometry; they are measuring numbers, numbers as magnitude. This is fine and good since one can live with two kinds of numbers as long as there is no inconsistency between them—that is, as long as the counting numbers can be identified with measuring numbers. This is the assumption of *commensurability*, the idea that all lengths have a common measure or that all ratios are represented by rational numbers. Unfortunately, this natural and aesthetically pleasing assumption was derailed by the theorem of Pythagoras and the argument that showed that the square root of two, the length of the hypotenuse of an isosceles right-angled triangle with sides of unit length, has standing as magnitude but not as multitude. The Pythagoreans believed that the world is number and they have bequeathed that amazing idea—that the world stands on a mathematical foundation—to us. Yet that foundation was not secure because numbers were ambiguous—they came in these two incompatible varieties, counting numbers and measuring numbers. The result was a crisis that called into question the very definition of number. The hypothesis of commensurability was used by Euclid to develop the theory of ratio and proportion that is central to much of his geometry. It is even conjectured[4] that the assumption of commensurability was used in the initial proof of the Pythagorean theorem. Non-commensurability, which should be thought of as reflecting an incompatibility that lies at the heart of quantity, now threw into question these pillars of Greek geometry.

Does multitude imply magnitude or vice versa? Does the continuous arise from the discrete or does the discrete come from an approximation of the continuous? This leads to a problem that I discuss in the following and explore in further detail in chapter 9. It is a deep problem, which is still not completely resolved. There are these two points of view. It will not do to say that one is right and one is wrong. The problem of quantity is not resolved by picking one approach over the other.

The tension between these two overlapping points of view is resolved through an act of creativity, which results, in this case, in generalized conceptions of quantity and thus generalized ideas of number. Without the tension, and without the incompatibility, the possibility of the creative resolution is removed. A good deal of the richness of the number concept arises from resolutions of the incompatibility inherent in quantity. Notice that I wrote resolutions and not resolution; there is, in general, no unique resolution to this kind of problematic situation. If the real numbers are a resolution, then so are the complex numbers,[5] various number systems that contain infinitesimals, and various other number systems that have been constructed by mathematicians. No ultimate and final resolution exists to the tensions inherent in quantity. It is and will always be a primal source of new mathematics, new science, and new ways of understanding the world.

Quantity as the Continuous
and as the Discrete

The ambiguity inherent in quantity finds expression in different conceptions of number. Today, we might think of the integers (and possibly the
rational numbers) as a model of multitude and the real numbers as representing magnitude. If we look at things in that way, then it is interesting that in one sense magnitude cannot be measured, since measurement
involves assigning a rational number to a given situation. Things like
duration, length, area, and volume are continuous and what measurement does is reduce the continuous to the discrete. This is also what a
computer does—it reduces the continuous to the finite, ultimately to 0s
and 1s. This model of the computer and of computation leads to a point
of view that holds that everything is finite, everything is discrete, that
the continuous is secondary and essentially a way of modeling the discrete. It is the discrete that is real. Ironically, in this view it is the rational
numbers not the real numbers that are "real." This is another way of
looking at the relationship between measurement, quantity, and number.
Here, measurement reduces magnitude to multitude, where quantity as
magnitude refers to the theoretical model, and quantity as multitude
refers to the practical results of the measurement process. One might say
that experimental science only deals with multitude, and that mathematics and theoretical science deal with magnitude.

A person can make a case that either aspect of quantity is the more
basic and that the other aspect is derived from it, but that is not what I
am saying. Mathematics and science deal with both multitude and magnitude. They are both ways of accessing the proto-concept we call quantity but that inevitably remains incomplete and partially inaccessible.
Quantity emerges into science as either multitude or magnitude, but this
is neither a passive either/or relationship nor can it be described by both/
and. It can only be called an ambiguity and, as such, is a relationship that
is generative and capable of further development.

To a large extent, the previous discussion goes backward and not forward. It starts with "number" and goes backward to "quantity." When
we reconstruct this process, we might well start with quantity as our
primordial concept and then define two varieties of the quantitative, af-

terward moving on to numbers that correspond to these two aspects of quantity. The first would lead to counting numbers, integers, and rational numbers, and the second to the geometric numbers. Later, we would discuss attempts to consolidate these two streams. However, that is not exactly what I did. I started with where we are now and looked for the origins of our present ideas and structures. A unidirectional flow of ideas is at best a reconstruction. It is useful and interesting but it misses something. It inevitably takes the present situation to be definitive. It tends to show how our present knowledge is superior in every way to the knowledge of the Greeks, for example. In doing so, it ignores the possibility that the Greeks knew things we do not know, that we have forgotten or suppressed. It seems heretical to suggest, but is nevertheless conceivable, that the Greek conception of quantity was in a certain way richer than our own, that their conception of number was deeper than ours. It was richer in the sense that a metaphor can be rich—because it comes with a large set of connoted meanings. It may well be that historical progress in mathematics is in part due to the process of abstraction, which inevitably involves narrowing the focus of attention to precisely those properties of the situation that one finds most immediately relevant. This is the way I shall view the history of mathematics and science—as a process of continual development that involves gain and loss, not as the triumphant march toward some final and ultimate theory.

Quantity vs. Quality

Notwithstanding this brief discussion of quantity and its role, I hasten to add that things are not as simple as they seem. Let us go backward again and note that the idea of quantity itself is usually set against that of quality. Quality refers to the essential identifying nature or character of something, but the nature of quality is quite elusive. This is the point that Pirsig makes in the statement that was quoted on page 4. In fact, his point is that *quality* is one of those ideas that cannot be captured definitively, just like infinity or randomness.

Let's start with an elementary example. Think about the number three. Is the essential identifying nature of three identical to its quantity? For the Pythagoreans, three was "the number of harmony, since it contains unity (the number 1) and division (the number 2)."[6] Is the harmony that the Pythagoreans discerned in "three" a quantitative phenomenon or is there something else going on? The modern response is that a number is by definition quantity, that quantity is identical to number, and any other aspect of "number" would necessarily be derived from the quantitative. However, if we think about number as a proto-concept, then we must allow for the existence of a larger set of references for the idea of number. This discussion of quality is an attempt to think about what an expanded meaning for number might be.

If we do some reading on the history of mathematics, we cannot fail but be impressed by how meaningful "number" was to so many different cultures. The idea that "the world is number" is a very powerful insight into some very profound aspect of reality. Number reveals the inner structure of a huge variety of things, not only things that are of interest to science but also to art, music, and religion. Modernity has reduced number to the quantitative, and we attempt to derive the qualitative from the quantitative. It is hard to remember that things were not always like this. Number was once deeper and richer than it is today.

The Pythagoreans had a way of thinking about the positive integers that sounds positively quaint to our modern ears. Three was the first male number, the number of harmony. Four was the number of justice and order. Ten, the *tetractys*, was revered because it represented the cosmos as a whole. Each number had its unique characteristics that echoed or revealed aspects of reality. Their numbers were richer than ours but it is difficult to understand what they were getting at, and so the modern mind tends to call such thought "numerical mysticism" in order to dispense with it. Yet the past is replete with such uses of mathematics. Davis and Hersh might have been talking about these qualitative aspects of number when they said, "We who are heirs to three recent centuries of scientific development can hardly imagine a state of mind in which many mathematical objects were regarded as symbols of spiritual truths or episodes in sacred history. Yet, unless we make this effort of imagination, a fraction of the history of mathematics is incomprehensible."[7] These people of the past were as intelligent as we are. What could they have been thinking of?

Sometimes, one gets an inkling of what the positive integers might have meant to the Pythagoreans when one reads accounts of mathematical savants. I am reminded of the famous story of the great self-taught

Indian mathematician, Srinivasa Ramanujan (1887–1920). At the time of this story, Ramanujan, who had been invited to England by the Cambridge mathematician G. H. Hardy, was in the hospital and Hardy had come to visit him. Ramanujan asked Hardy, "What was the number of your taxi?" Hardy responded, "1729, a most uninteresting number." "On the contrary," says Ramanujan, "it is the smallest number that can be written as the sum of two cubes in two different ways."[8] What an intimate knowledge of the positive integers Ramanujan reveals here! For him, the natural numbers were not just an infinitely long sequence with each number one larger than the one preceding it. Each integer was unique to him. It had its own characteristics and properties, its own relationship to other numbers, its own meaning.

Savants seem to experience numbers differently than the rest of us. It is true they calculate faster, but rapid calculation is just one part of what they do. The key to their performance is that they seem to have a different experience of number. They don't even calculate the way the rest of us do—they seem to have the capacity to "know" numbers in a way that is qualitatively different. The fascinating story of Daniel Tammet offers us a window. Because he is both highly intelligent and has exceptional savant abilities, his words give us an unusual glimpse into the way his extraordinary mind works. And it works in a very different way than most people's. In his biography, he articulates his experience of numbers. "I see numbers as shapes, colors, textures, and motions. The number one, for example, is a brilliant and bright white, like someone shining a torch beam into my eyes. Five is a clap of thunder or the sound of waves crashing against rocks." He says, "Scientists call my visual, emotional experience of numbers synaesthesia, a rare neurological mixing of the senses." He goes on to describe the way he multiplies 53 by 131. "I see both numbers as a unique shape and locate each spatially opposite the other. The space created between the two shapes creates a third, which I perceive as a new number: 6,943, the solution."[9]

Tammet's experience of the integers is qualitatively different than our own. He claims it is his synaesthetic way of experiencing numbers that accounts for his extraordinary feats of calculation and memory and this seems consistent with other accounts of people with extraordinary memories.[10] What is interesting is the possibility that numerical synaesthesia is not the byproduct of a damaged mind (Tammet is an epileptic) but that some condition of Tammet's brain allowed him to become conscious of connections that most people either do not have or are not aware of. Are savant abilities, like synaesthesia, a defect or a gift? Why don't we all have these abilities? Actually, this question was asked by the researchers Allan Snyder and D. J. Mitchell, whose work is discussed in the book *Musicophilia* by neurologist Oliver Sacks. "They suggested that the mechanism for such skills might reside in all of us in early life but that as the brain matures, they are inhibited, at least from conscious awareness. They theorized that savants might have 'privileged access to lower levels of information not available through introspection.'"

Suppose that we all have the latent ability to experience numbers and other scientific concepts in a multi-dimensional, multi-sensory way and that these abilities could be enhanced through training. This would lead to different ways to think about scientific problems and possibly to original approaches to these problems. Creative individuals often have a way of thinking about scientific situations that are unique. Perhaps we shall get some insight into the savant's experience of numbers by considering another set of savants, this time people with limited intellect.

The next story I shall briefly describe forms a chapter in another book by Sacks, the fascinating bestseller *The Man Who Mistook His Wife for a Hat*. He writes about some patients of his, twin bothers John and Michael,[11] who were idiot savants, and had been variously diagnosed as autistic, psychotic, or severely retarded. They had the "usual" savant abilities for rapid and uncanny "calendar computations" and for remembering lengthy lists of digits. Yet "They cannot do simple addition and subtraction with any accuracy and cannot even comprehend what multiplication or division means." They had IQs of sixty. On the other hand, they could instantaneously factor 111 into 37×3 and determine whether or not huge numbers (five, six, or seven digits) were primes (had no divisors).

How did they work out the answers to these difficult problems? This was the question Sacks asked the brothers. But they indicated to him that "they did not work it out, but just saw it, in a flash." "Is it possible," Sacks said to himself, "that they can somehow 'see' the properties, not in a conceptual, abstract way, but as *qualities*, felt, sensuous, in some immediate concrete way?" It is this very aspect of quality in number that we are investigating, and Sack's case history allows us to differentiate quality from quantity.

Sacks goes on to say,

> The twins, I believe, have not just a strange "faculty"—but a sensibility, a harmonic sensibility, perhaps allied to that of music. One might speak of it, very naturally, as a "Pythagorean" sensibility. . . . One's soul is "harmonical" whatever one's IQ, and perhaps the need to find or to feel some ultimate harmony or order is a universal of mind, whatever its powers, and whatever form it takes. . . . Mathematicians have always felt number as the great mystery, and the world as organized, mysteriously, by the power of number.[12]

"Organized, mysteriously by the power of number," he says, and you could say the same thing about the power of mathematics as a whole or the power of science. But if we talk of "power," then we must conclude that the nature of mathematics and science is more that of a dynamic process as opposed to something that is static and unchanging. This is the reason it is better to talk about "doing mathematics" or "doing science" as opposed to just talking *about* mathematics and science.

Sacks wrote of a "Pythagorean sensibility." Could it be that the case histories of idiot savants, precisely because of their low IQs, allow us to isolate this Pythagorean sensibility, the qualitative aspect of the mind that we can perhaps begin to see as another legitimate type of intelligence, one more rudimentary but more immediate than the quantitative and the analytic? Perhaps this will give us a glimpse into the mystery of number and its potential to organize the world.

Sacks continues,

> Hermann von Helmholtz, speaking of musical perception, says that though compound tones can be analyzed, and broken down into their components, they are normally heard as qualities, unique qualities of tone, indivisible wholes. He speaks here of a "synthetic perception" which transcends analysis, and is the unanalyzable essence of musical sense. He compares tones to faces, and speculates that we may recognize them in somewhat the same personal way. In brief, he half suggests that musical tones, and certain tunes, *are*, in fact, "faces" for

the ear, and are recognized, felt, immediately as "persons," a recognition involving warmth, emotion, personal relation.

So it seems to be with those who love numbers. These too become recognizable as such—in a single, intuitive, personal "I know you!"

Sacks provides us with another hint about how to make sense of quality and differentiate it from quantity. Quality arises from a "synthetic perception" as opposed to an analytic one. A synthesis is defined as "a new unified whole resulting from the combination of different ideas or objects," but this definition does not quite capture the way in which I shall use the word. Synthesis involves seeing the situation as a whole. When you do this, the components fade into the background and the "whole" becomes the foreground. Such perception is not constructive, you don't reason your way to it; it is not a question of the organization or reorganization of a number of predetermined elements. You "see" it in a moment of recognition. It's like when you forget someone's phone number. It doesn't come back to you digit by digit. At some level you know the number since you can clearly say that a particular guess is wrong. But somehow you just cannot access that knowledge at a given moment in time. However, when it comes back to you, it all comes back at the same time, it is all one number. Further, when it does come back, you *know* without a doubt that it is the right number. The gestalt psychologists studied synthetic perception extensively. One of the most notable of these scientists, Max Wertheimer, said in a talk in 1924,

> The fundamental "formula" of Gestalt theory might be expressed in this way: "There are wholes, the behavior of which is not determined by that of their individual elements, but where the part-processes are themselves determined by the intrinsic nature of the whole."

You might say that for the twins, number was a "gestalt," a whole. This provides us with another meaning for the expression "a *whole* number"—every positive integer is first and foremost a whole. It registers in the mind like those faces of Helmholtz's—a complex pattern that is dynamic and instantly recognizable.

Synthesis is a way of seeing "wholes."

Analysis, on the other hand, involves taking wholes apart to investigate their constituent elements. Roughly speaking, we might speculate that IQ measures the potential for analysis but that the potential for synthesis is (to a certain extent) an independent ability. Of the two, synthesis is more elementary, as evidenced by the story of the twins. It probably goes on in a different region of the brain. The capacity to know number, in a rudimentary way at least, is very primitive, and so is "quality." It is now well established that a propensity for numerosity—what is called subitizing, or *knowing without counting* the cardinality of small collections of objects—is common in the animal kingdom and is a capacity that human infants give evidence of long before they learn to speak.[13] Almost by definition, subitizing is synthetic not analytic. Some people claim that "number" is conceptual, and so subitizing does not involve number as much as it involves "numerosity" or the potential for number. This fits in well with what I am saying. The synthetic is another way of knowing, of using the mind. Actual mathematical practice involves the integration of these two features—the analytic and the synthetic, the quantitative and the qualitative. This may be the reason that people love areas of mathematics like Euclidean geometry that integrate these two aspects of the mind so wonderfully.

The Qualitative Properties of Non-integers

So far, we have focused on the qualitative properties of the counting numbers. But what about other kinds of numbers? Do they also have properties we could call qualitative? Perhaps thinking of them as quality in addition to quantity may help us think about those mysterious constants that arise in mathematics and its applications. I earlier discussed Chaitin's number, Ω, from a qualitative point of view, and we could also think of numbers like π, ϕ (the golden mean),[14] e,[15] and even $\sqrt{-1}$[16] in this way. Each one has a significance that goes beyond their numerical value. Each one tells us something profound about the world we live in.

Take π, for example. It is a totally different kind of animal than the integers: 1,2,3, ... If we happen upon it with only a knowledge of the rational numbers, we might even ask whether it is a number at all! It's decimal representation is so complex that many questions about it are still far from being resolved.[17] One of the deeper aspects of π is that it is the constant that helps you translate from the linear world of straight lines to the periodic world of circles, For example, there is the well-known formula $C = \pi d$, that relates the length of the circumference of a circle to its diameter. The quantitative way of looking at this equation is that it is just a relation between three numbers. A more qualitative way of looking at it would be as a metaphor, analogous to how I discussed Einstein's equation in chapter 5. We would then recognize that the linear and the periodic are two ways of looking at the natural world. This is reflected in the fact that we all have two biological systems for measuring time. Linear biological time is called "interval timing" and consists of the ability to tell elapsed time. On the other hand, there is what is often called the "biological clock" or "circadian rhythm," a roughly twenty-four-hour cycle in the biological, physiological, or behavioral processes of living beings. Each of these biological systems has given rise to a distinct way of looking at the world. π would then represent the key to comparing the linear world with the non-linear and periodic one, of translating from one world to the other.

The Neurological Correlates of the Analytic and the Synthetic

In recent years, there has been a great deal of research revolving around hemispheric differentiation in the brain, the differences between the right hemisphere and the left hemisphere. A most moving and fascinating first-person account of the difference comes from Jill Bolte Taylor in her recent book and lectures (one of which is available on YouTube).[18] Dr. Taylor is a brain researcher who had a stroke in the left half of her brain. She survived and courageously fought her way back to health and has devoted a great deal of her time to communicating her experience, letting the rest of us know "how it feels to have a stroke."

She says that

> Our right hemisphere is all about the present moment.... It thinks in pictures and it learns kinesthetically through the movement of our bodies. Information in the form of energy streams in simultaneously through all of our sensory systems. And it explodes

into this enormous collage of what this present moment looks like. What this present moment smells like and tastes like, what if feels like and what it sounds like.

My left hemisphere is a very different place. [It] thinks linearly and methodically ... [it] is designed to take that enormous collage of the present moment and start picking details and more details and more details about those details. It then categorizes and organizes all that information.... And our left hemisphere thinks in language.

Because of her stroke, Taylor lost her left hemisphere, it went "offline" to use her words. As a result, she was able to give us her fascinating account of this other kind of intelligence that we all possess. We have, she says, two cognitive minds. The important question, the right question, is how these two minds collaborate to produce science. The wrong question is how the analytic mind produces science.

Normally, we think about science and mathematics as though the left brain, the analytic intelligence, is all that is happening: taking things apart to discover their most elementary properties and then reassembling these "atoms" or "axioms" into complex structures. We normally think of mathematics and therefore of numbers as exclusively quantitative. As a result, we imagine that the role of mathematics is to "reduce" things to the quantitative. Somehow, the analytic and thus the quantitative seem more "certain" than the qualitative and the synthetic. The qualitative seems to be more mysterious, less certain in one way, but more immediate in another. In our quest for certainty and our resulting urgency to reduce everything to the quantitative, we sometimes forget that the whole edifice of science and mathematics stands on a foundation of the qualitative. Without the synthetic we could not see "wholes" or "units" at all; the process of science could not even get started.

These two legitimate points of view—these two different intelligences—are at play in every human being. We are, each one of us, split down the middle, so to speak. We are two different people, two separate intelligences. It is true that these two intelligences communicate with one another. However, it is a peculiar kind of communication when speech is localized in the left hemisphere. From the point of view of analytic intelligence, the non-verbal right hemisphere is inevitably a mystery. It is inaccessible if accessibility implies the capability of being reduced to words and analysis.

CONCLUSION

This chapter gives just a taste of the profundity of the idea of number. We saw how number should be thought of as a proto-concept with an unlimited potential to appear in new and original ways. In a recent paper, Philip J. Davis makes a statement about mathematical problems that applies well to the present discussion of number, "meaning is dynamic and ongoing and there is no finality in the creation, formulation and solution to problems, despite our constant efforts to create order in the world."[19] The meaning of number is also "dynamic and ongoing" with "no finality." The dynamism in the case of number comes from an unresolved tension within the idea of number that is associated with the various polarities that were presented: magnitude and multitude, the geometric and the algebraic, the continuous and the discrete. Number is the essence of quantity and is therefore at the heart of various attempts to quantify phenomena. Just like the world of the Pythagoreans, the modern world also stands on a foundation of number, but with a qualitative dimension that has been lost, or at least substantially diminished. As a result, our culture has been impoverished.

Since the qualitative is related to the synthetic, and the quantitative to the analytic, the polarity of the qualitative and the quantitative is reflected in the way we think and interact with the world. In this way, number is an idea that stands at the very heart of any systematic study of both the human mind and the natural world.

8

Science as the Ambiguous Search for Unity

Man knows himself only to the extent that he knows the
world; he becomes aware of himself only within the world,
and aware of the world only within himself.
—Johann Wolfgang von Goethe

INTRODUCTION

This is a key chapter, but a difficult one. The difficulty is precisely the
central paradox of this book: If there is indeed a blind spot, how do you
talk about it? If this uncertainty is present in all of our concepts and theories, does this mean we mean must refrain from talking and thinking?
Of course not! But it does mean that our theoretical thinking should
be put within a larger perspective, a perspective that is open-ended. In
particular, certain ideas transcend any particular definition. This was the
way in which number was discussed in the last chapter. I shall use the
word transcendental for this way of looking at ideas, as a pointer to the
reality that lies beyond any particular formulation.

   This chapter is about unity in general and, in particular, the unity
between human beings and the natural world that Goethe alluded to in
the opening quote. It owes a great deal to the discussion in chapter 5 of
Low's *The Butterfly's Dream*. Unity, like randomness, infinity, or ambiguity is a transcendental idea. When you talk about an idea from the
transcendental dimension, you can't begin in the usual way by giving
an exact definition, because circumscribing it in this way would strip it
of its transcendental quality. One must proceed, as I did with ambiguity,

by introducing it in a variety of ways, by discussing a collection of its connoted meanings, with the hope that the reader will intuit the larger meaning. This is why discussing unity may be frustrating to some readers—those who demand that the only way to proceed with this kind of discussion is to begin by defining all of your terms. It is the same reason why ambiguity may be hard to grasp, and why students have difficulty with the ideas of infinity and randomness. I want to encourage readers not to give up on this prematurely—I also find these kinds of ideas difficult. They can only be understood in a limited way but they are real!

The problem in this chapter revolves around the description of the world that is associated with the philosopher René Descartes. Our conscious experience tells us that Descartes was right: there is a world of things and a world of the mind, an objective and a subjective world. Unfortunately, Descartes was also wrong—reality is monistic, even though experience is inevitably dualistic. This means that the objective and the subjective are united at a higher, transcendent level. This level is what I shall call unity, and this chapter is my attempt to describe the nature of this unity. Unity is not just a universal container like the "set of all sets." It is both local and global, process and object, subject and object. To use a simplistic metaphor for the unity of reality: It is like a glove; the objective might be seen as the outside, and the subjective as the inside, but such distinctions are artificial since basically there is just one glove.

Goethe is talking about this unity in the chapter's opening quote, that self-awareness cannot be separated from knowledge of the world. The philosopher Baruch Spinoza, a favorite of Einstein, said, "The greatest good is the knowledge of the union which the mind has with the whole of nature." They are both talking about the unity that stands at the origins of the entire scientific enterprise. Unity is the mystery that lurks behind philosophy's famous mind/body problem. It is the very essence of science, as it is of any serious attempt to describe the natural world and the human condition. How extraordinary it is that our scientific minds can penetrate so deeply into the profundities of nature! How wonderful it is that we are one in this way with the natural world. Unity lies behind Einstein's "experience of the mysterious" that we discussed earlier, and it is the theme of this chapter.

UNITY AS HARMONY

Unity has already been alluded to at various places in this book. For example, in the last chapter, the analytic was contrasted with the synthetic as two cognitive modes to which human beings have access. The synthetic involves seeing things as "wholes" or "unities," as opposed to seeing them as amalgams of their constituent elements. It is the synthetic mode of being, the essential insight of gestalt theory, which leads to the subject of this chapter. We all possess an entire cognitive system whose function is to perceive unity. Unfortunately, unity cannot speak, so its existence must be inferred indirectly.

Unity is a basic intuition we all share. We have the intuition that there exists a state of harmony that we can discern both in ourselves, in the natural world, and ideally in our relations with the world. Despite the dichotomies that so characterize human existence, we do not experience ourselves as divided, as multiple independent tendencies, but as one unified and harmonious being. We may have two different cognitive modes, but in a well-functioning individual, like a good researcher, this whole functions like a well-oiled machine. No one could possibly deny the sense that they are one and not merely various discordant tendencies. It is against this background of unity that all the conflicts of life play themselves out.

But unity does not just refer to the unity of the personality. Everyone, at some point in his or her life, has had a so-called "peak experience" in which they sensed an extensive harmony between themselves, other people, and the whole of nature. John Lennon[1] hoped that one day, "all the world will be as one," but the oneness he hoped for does not exist in some utopian future. It exists in the present moment. In the words of an anonymous woman of thirty-eight:

> I was standing alone on the edge of a low cliff overlooking a small valley leading to the sea…. Suddenly my mind "felt" as though it had changed gear…. I still saw the birds and everything around me but instead of standing looking at them, I *was* them and they were me. I was also the sea and the sound of the sea and the grass and sky. Everything and I were the same, all one.[2]

This is the experience of unity. Unity finds its way into science in the feeling that the universe possesses an underlying harmony, what has been called "the music of the spheres." It is a sense that you get most powerfully in the thought of the ancient Greeks but it also pervades all of science. We have the conviction that things make sense, that the natural world *is* comprehensible. This feeling for the natural harmony that pervades the universe and ties human intelligence to it is an aspect of Einstein's "cosmic religious feeling." It is equally Spinoza's, "union which the mind has with the whole of nature" that we mentioned earlier. We have the profound intuition that we are one unified being, and—moreover—so is the natural world. Of course, we can understand the universe because we are made of the same stuff as the universe—we are one with the world. It would be a big mistake to doubt the intuition of unity. Nevertheless, unity itself remains opaque. What is it really?

Let me try to bring the discussion of unity down to earth by talking about it in a different way. When I am writing down these words and ideas, I begin with a topic and develop a whole set of related ideas, facts, concepts, and quotations. At some point, I start writing, but things invariably do not fit together—the distinct elements do not add up to anything coherent. My experience is that if I stay with the situation there often comes a moment when things begin to come together, when I say to myself, "this doesn't sound too bad, I think I'll stick with this." This moment marks the appearance of unity: unity as a chapter, a section, a paragraph, or only a sentence. This is also how it is when one does science or any other creative activity. After a lot of hard work, if you are lucky, unity appears in the guise of an idea that pulls things together and makes sense of the situation. But unity is not merely the finished product—the chapter or the theorem or the research result—unity is also the process. You learn to have faith in the process. If you struggle with the situation in an authentic way, think about it, play with it, remain concentrated on it for as long as it takes, then it will come together. Unity is the process, but unity is equally the result of the process. It is interesting in this regard that I cannot force the writing to sound right, forcing is not the way to get unity to appear—all that can be said is that at some stage it sounds right to my ears and to my mind and I then merely acknowledge what has happened. This is an ordinary run-of-the-mill transcendent experience, which involves going beyond the duality of the subjective versus the objective. It marks the appearance of unity.

Unity as the Appearance of Order

The previous paragraph gives us another way to think about the elusive notion of unity. Unity can be understood in terms of the appearance of order and structure. Order involves an internal coherence, coherence being another word for unity. However, we usually think about order as a phenomenon that lies within some objective situation. For example, that gestalt picture I discussed in chapter 4 is ordered when we see it as a young woman, since viewing it in this way unifies all of the elements into one coherent picture. So it is with much of science. A scientific "law" is the observation of a certain pattern or regularity and thus involves the discernment of unity.

We usually think about order as existing in the objective situation. The picture *is* the young woman, for example. We don't attend to the fact that we perceive order, that we bring order to a situation through attention and intelligence. Order, you might say, is also in the eye of the beholder; it must be perceived. We order the night sky by creating pictures that we call constellations. The stars that make up the Big Dipper are really out there, but the Big Dipper is something that we see. This is the unity of the human mind with nature that I discussed at the outset of the chapter.

If we think of unity as dynamic rather than static, we can understand why unity is connected to the appearance of order. Unity is not merely an objective order; it is a force, a tendency toward order. This should not be so surprising for, after all, isn't this what the theory of evolution is telling us?

Science is concerned with revealing order. In complexity theory, it has been noted that order emerges spontaneously from certain kinds of complex situations. It is unnecessary to posit some outside agent who creates order. The universe, according to many contemporary scientists, is self-generating. A tendency toward order is built into the natural world. This tendency is part of what I am calling unity.

Now we might think that order is something self-evident—that we would all recognize an ordered situation as opposed to a disordered one. However, it turns out that order is quite subtle. For example, randomness, which is ostensibly the lack of order, can actually be a source of order in various ways. One is the way in which a random coin flip will have a fifty percent chance of being either heads or tails. Thus statistical order, the fact that the possible outcomes are equally probable in the long term, is generated by randomness. In the theory of evolution, the random mutation of genes is the motor that drives evolutionary development, the order that we observe in the biological world.

Order and randomness are related to certainty and uncertainty and, as such, are connected to the main themes of this book. We tend to identify unity with order and certainty, yet without uncertainty there would be no context within which unity could emerge. Unity lives at a deeper level than this dichotomy.

Order and randomness are notoriously difficult to define because, at a most basic level, they are not concepts at all but proto-concepts. Order is not objective because, like the constellations of the stars in the sky, it requires the existence of an observer or at least a point of view. Nevertheless, you cannot say that order is arbitrary. Clearly some patterns run deeper than others. You cannot arbitrarily impose some artificial pattern on data and think you are doing science. The patterns that last can be used to predict the results of future experiments; they are objective.

It is interesting to speculate that the emergence of order has something to do with intelligence, whether or not the situation we are discussing involves human beings. The universe in its early stages was characterized by the formation of hydrogen, an early example of the emergence of order. Human life at every level, from the structure of the physical body to the structure of the mind involves the emergence of an order that is incredibly complex. It is not only that the human mind creates order but also that the human mind is the result of the tendency toward the production of order that is at work in all places and at all times. When I use the word unity, it is this tendency that I have in mind, not just the objects or beings that result from this tendency. It is a dynamic force that pushes us all in the direction of meaning and significance. In anyone who is sensitive to it, and this includes every creative scientist and artist, it feels more like an imperative: *There must be one!* Unity demands to be expressed. When researchers in complexity theory, for example, refer to "emergence," what they are really saying is that they have rediscovered the force of unity.

UNITY AS THE ONE: THE SOURCE OF MATHEMATICS

Hear, Israel, the Lord is our God, the Lord is One.
Deuteronomy 6:4-9

In this section, I want to pursue the way unity finds its way into mathematics and science. I shall begin by discussing the number one, the origin of mathematics and of science. "One" marks the emergence of unity into the conceptual world.

To get a handle on the mysterious but related ideas of "unity" and the "One," let's go back to Plotinus (204–270 C.E.), one of the most influential philosophers of antiquity after Plato and Aristotle. The One is one of the three basic principles of Plotinus's metaphysics. For Plotinus, the One is not a number. It is "the absolutely simple first principle of all. It is both 'self-caused' and the cause of being for everything else in the universe." Plotinus says:

> It is by the One that all beings are beings. If not a One a thing is not. No army, no choir, no flock exists except that it be One. No house, nor even ship except that it exists as the One.[3]

The One, for Plotinus, is not a concept. In fact, it cannot be described directly and explicitly. I made the same claim for unity, and as near as I can make out, the One for Plotinus *is* what I've been calling unity. It refers to something that is not yet conceptual. It is a generator of concepts, or what the last chapter referred to as a "proto-concept." In fact, for the purposes of discussing science, the word "One" has more resonance than the word "unity." The world appears to us, not as an undifferentiated soup, not as chaos, but as being made up of objects, of units or unities. I look out my window and see a tree as a unit. Thus, unity does not only refer to "the state of being united or joined as a whole"[4] in the sense that any finite collection of separate things—for example, a couple—is, or can be thought of, as a unit (or unity). It also means that every object is *one* object; thus, every object, to be an object, must be a unit.

Every scientific situation has two elements: objects such as atoms, molecules, or people; and relations (or forces) between these objects. In mathematics, for example, we have sets and functions. I have just said that each of the objects (but also any set of objects) is an instance of (the principle of) the One. However, it is equally true that relations, by tying the objects together, are also an example of the same principle; relations create a structured unity. If we look at the positive integers, then there is the following relation—two integers m and n are related if their difference, $m-n$, is divisible by two. In this way, all even numbers are related to one another, as are all odd numbers. This isolates new unities, namely the odd and even numbers.

Oneness or unity is so basic that we take it for granted. However, it is easy to see that without such a principle it would be impossible to even get started on the scientific enterprise. One way of seeing this is to imagine teaching this principle to a computer. How can a computer that is receiving data from a visual sensor break down the contents of my desktop into distinct objects—into books, papers, pens, telephones, computers, and so on? For the computer, everything that is sensed is just data—in other words, chaos. It requires another principle to convert this chaos into a structured unity. For one thing, it requires a form of intelligent discrimination. Maybe this is another way to understand the passage from Genesis that begins, "The earth was without form and void, and darkness was upon the face of the deep." It was out of this formless void, through, the principle of the One, as Plotinus might say, that the world as we experience it came to be.

The Conflicting Faces of the One

The One is ambiguous—both unitary and dual. That something can be at the same time one and two, that unity is divided within itself, is hard to get your head around. In a way, it is the fundamental mystery. How can something be both unified and divided at the same time? This is a wonderful question that pushes us in the direction of movement, of creativity, and of evolution, which is essentially just another form of creativity. When we begin to think of the One, we think of it as something static and objective. But the One is neither static nor objective. On the contrary, it is dynamic and generates change. The dynamism we observe in the world and in ourselves has its origins in this paradox of the One that is Two, and yet cannot be Two because it is One. The duality that resides in ambiguity is not acceptable because there exists an imperative to return to the One. It is this impossible situation that drives us to look for creative resolutions to problems, to do research. Yet when we are successful and the One appears to us in the guise of a resolution of the problem that we are working on, that is not the end of the story. In fact, the story has no end because conflict in the form of disunity will appear again, and so a new cycle will begin.

This conflict within the One is clear even from our ordinary use of the term. "One" is the essence of the simple and the minimal. As a process, it involves breaking things down as far as they can go—into cells, or molecules, atoms, or subatomic particles, depending on your field of interest. Using "one" in this way, saying one pen or one electron, involves looking at matters from the outside, from the observational point of view. Low has called this the "exclusive" one. In mathematics, this "exclusive one" appears as the point in geometry, the element in set theory, or simply as the number 1. It is the simplest constituent of a given situation. The analytic method, so prevalent in science, consists in breaking a given situation down into its simplest and most elementary constituents, into units, and then reconstituting the system from that foundation.

However, the term "one" can also be used in an inclusive sense, as when we speak of a group of people "acting as one." To be one with others means to be part of a larger unit. Thus, "man and wife are one" means that these two people now form a single married couple. The inclusive one also comes with a point of view but in this case it is a view from inside. Thus, "one" can be used in these two conflicting ways—both as inclusive and as exclusive. Note that these mirror the way in which human consciousness is bifurcated into observer and participant. We can see matters from the inside or from the outside, and there is an aspect of "one" corresponding to either point of view. On the other hand, the one is singular by definition, so it is unacceptable for there to be two "ones," just as there is an intrinsic conflict between my "two cognitive intelligences." I am two, but I cannot be—I am one. The inclusive one cannot be distinct from the exclusive one since one is one and not two. This is the impossible situation that we face as human beings. How could it not find its way into science?

The Counting Numbers as the Unfolding of the One

In chapter 8, in the discussion of "number," I made the point that the Greeks' understanding of number was not identical to our own. In some obvious ways, their understanding was inferior; they never developed the real numbers—for example, where each number, 1/3, $\sqrt{2}$, or π, is considered to be an infinite decimal. In others ways, the Greek understanding of number may well have been deeper and richer than ours. Consider the Pythagorean understanding of the positive integers. We tend to think of these numbers as a collection, sometimes denoted by the symbol $\mathbf{N}$, in which each integer is essentially on an equal footing with every other. The quantitative properties of the integers are very subtle and the field of number theory is devoted to discovering these properties. Number theory is one of the most active fields of modern mathematical research. It has witnessed some of last century's most striking successes, such as the proof of Fermat's Last Theorem.[5] It also contains many problems such as Goldbach's conjecture,[6] which remain unresolved to the present day.

The Pythagoreans focused on properties of the integers that today we don't pay much attention to. In the spirit of the last chapter, I shall call these qualitative properties rather than quantitative properties. To begin with, one and two were not integers for the Greeks; positive integers began with three. Our discussion of "one" might give us some insight into this distinction. The integer one, as we now conceptualize it, is but a pallid reflection of the unnamable One of Plotinus. It was not as though there were integers of which the number one was a mere example, but that if not for the One there would be no integers at all.

One represents unity, but it is a complex unity that contains an inner division, a unity that is ambiguous. One represents that unity, but "two" also represents this unity, this time focusing on the division. Another way to see this is that "two" represents duality, which is an aspect of unity. Because unity is dynamic and not static, because there is a conflict between the "one" and the "two," the basic forces we are calling "one" and "two" generate the whole sequence of positive integers. This sequence begins with the number three. Three represents the emergence of a new equilibrium from the unstable situation that we could designate as one/two. But no equilibrium is permanent, and as it breaks down it reestablishes a higher-level equilibrium that is represented by the integer four, then another for five, and so on. Every integer is therefore a development of unity, a representation of structured unity.[7]

The number sequence is like a recapitulation of the development of the universe. This may be the reason why the Pythagoreans saw "number" as the underlying principle of the world. However, the development in this case should not be thought of as a temporal development. When Plotinus or the Rig Veda say that everything *starts* with the One, "start" does not refer to time since, from this point of view, time itself arises from the same source—the force of divided unity.

The Unreasonable Effectiveness of Mathematics

Nobel Prize–winner Eugene Wigner[8] raised the question of the effectiveness of mathematics in an influential and often quoted paper that addressed the central mystery about mathematics, namely: Why does mathematics work so well? Why is differential geometry so well suited to describing the general theory of relativity? Why does Hilbert space theory reveal the essence of quantum mechanics?

Historically, one of the most dramatic cases of this phenomenon was the discovery of the planet Neptune in 1846. As the historian of science, Daniel J. Cohen, says in his book,[9] "Neptune was the first heavenly body found by mathematical prediction. Without peering into the sky at all, Adams and Le Verrier[10] independently calculated the location of the planet through geometrical analysis and the laws of gravitation." This dramatic discovery caused a sensation because it seemed to confirm that the mathematical laws of science governed the universe or, more succinctly, that God was a mathematician.

Why this should be so was, and remains, unexplained. John D. Barrow pinpoints this issue at the beginning of his fascinating book on the nature of mathematics:

A mystery lurks beneath the magic carpet of science, something that scientists have not been telling, something too shocking to mention except in rather esoterically refined circles: that at the root of the success of twentieth century science there lies a deeply "religious" belief—a belief in an unseen and perfectly transcendental world that controls us in an unexplained way, yet upon which we seem to exert no influence whatsoever.[11]

According to Barrow, the correspondence between the unseen world of mathematics and the natural world is something we accept as an act of faith akin to the way some people accept the tenets of religious belief. The efficacy of mathematics is a mystery with no rational underpinning, yet it is a faith that is shared by mathematicians and scientists, not to mention businessmen, civil servants, and countless others. In fact, it is shared by anyone who uses mathematical or statistical arguments to support some proposed course of action. If culture means the existence of a shared world view, a shared set of beliefs, then the effectiveness of mathematics is one of our cultural axioms.

Many mathematicians are Platonists; they believe Platonism captures a genuine aspect of mathematics. For them, mathematics is real; it possesses an existence that is independent of its creators. And this is what it feels like to do mathematics. You might not know the answer to the problem you are working on, but you have faith that it exists, somewhere out there, just waiting for you to access it. In recent years, the eminent mathematician and physicist Roger Penrose has had the courage to go public, so to speak, with an unabashed affirmation of his Platonism. Because this belief seems so foreign to the usual way in which most people think about science, because it seems to be almost "religious" in nature rather than empirical, most people prefer to ignore it or pretend it is not a legitimate part of science. If it is not ignored, then it may be relegated to the psychological domain—but ultimately this is not very satisfying. Mathematics works too well to be arbitrary; there is something systematic going on in the mathematical description of reality.

Why not take the effectiveness of mathematics at face value? Why not accept it as real, and instead of asking why it is true, why not ask what can be inferred from the effectiveness of mathematics? For indeed the universality of mathematics and its evident utility in describing the natural world has something profound to teach us, something about unity, the unity between knowing and being, between the human mind and the natural world. In Descartes's famous division of the world into objects and mind (the famous body/mind duality), empirical science naturally appears to have a home in the former, while mathematics appears to live in the latter. Mathematics is effective precisely because these two domains are connected at a level that is deeper than either, because our minds are one with the natural world and are not only situated, as it sometimes appears, on the outside of the world looking in. Thus, Wigner's "Unreasonable Effectiveness of Mathematics" is pointing to a healing of the mind/body divide through mathematics. Mathematics, you could say, reveals that the world is one; it reveals unity.

The same lesson, though in a less direct way, is to be learned from Platonism in mathematics. It is not merely that the natural world is real and the world of mathematics is artificial. The world of mathematics is not merely subjective, a matter of opinion. It is solid and objective. Yet that objectivity is an objectivity of the mind. We should rethink the very notion of what we mean by objectivity and subjectivity. I mentioned in earlier chapters that subjectivity may be taken to mean something that is colored by prejudice and personal idiosyncratic opinion. However subjectivity may also refer to matters of the mind. If that is the case, then a mental phenomenon may also be objective. In fact, a basic rule of our mental universe, especially in mathematics, would seem to be the process of reification, wherein we make processes or patterns into objects. In arithmetic, this might involve taking the process of repeated addition $(2+2+2)$ for the product (3 times 2), but our language is full of instances where verbs are turned into nouns—for example, we talk of going for a "run" or sending an "e-mail." The objects of the world are constructed by us and solidified by social convention. This does not mean they are imaginary, but that the objective and the subjective are not disjoint categories, and instead collaborate to produce what we call reality. The lesson of Platonism is that the world of the mind is real, a lesson that has been rediscovered by certain cognitive psychologists in their study of metaphor.[12]

Discovery or Invention

One of the most long-standing questions about mathematics is whether it is discovered or invented. Platonism takes the stance that it is discovered; formalism, which sees the axiomatic method as fundamental to math, stands more or less on the side of invention. It is not necessary to make an absolute choice since many people have argued that mathematics involves a combination of discovery and invention. For example, Reuben Hersh[13] places mathematics in a socio-historical setting in which human beings invent the mathematical system (like the counting integers), but having done so, properties of the system (like Fermat's Last Theorem) are objective phenomena to be discovered. This last way of looking at mathematics is very persuasive, but I shall contribute a slightly different perspective that is based on the discussion of unity.

The suggestion is that we treat the conundrum of discovery versus invention in mathematics in the way we thought about the unreasonable effectiveness of mathematics. Discovery and invention are two equally valid ways to approach mathematics. The perspective of "discovery" tells us that mathematics exists in some domain that is objective. "Invention" emphasizes that mathematics resides in the domain of the mind. These two approaches conflict if we believe that a strict duality exists between these two domains. The "unity" perspective that I have introduced in this chapter would say that the ambiguous situation of discovery/invention has a resolution at a higher level. The existence of these two perspectives on mathematics can be looked upon as evidence for this higher-level unification. Mathematics is one discipline but this discipline comes with a dual perspective. Mathematics itself is evidence for this unity that I have been discussing.

Now, the preceding discussion not only applies to mathematics but also to science. Here, the domain of discovery would be the natural world, the world of data and experiment. The domain of invention is the theoretical world. Thus, Einstein's "cosmic religious feeling" stems from the realization that these two dimensions are ultimately One, the sense of wonder that accompanies the realization that the mind can penetrate so deeply into the secrets of the natural world. The reason that this is possible is because mind and matter are two perspectives on one unified reality. In other words, science itself is evidence for the existence of unity on the deepest level, a unity that has this dual perspective that we can call mind and matter, or simply knowing and being.

Neuroplasticity and the Mind/Body Problem

Evidence for the complex unity of mind and nature comes from the relatively recent discovery of neuroplasticity.[14] Neuroplasticity concerns itself with the complex unity of the brain and mind. Until recently, the conventional scientific view was that mind could be reduced to brain, that the physical brain was the primary phenomenon and that the mind was merely an epiphenomenon. Yet in recent years, evidence has emerged that the physical configuration of the brain is malleable and can change as a result of learning, thinking, and other mental activities—in short, that the mind can influence the brain. This evens up the scales between brain and mind. Thinking of one as primary and the other as secondary does not seem to reflect what is actually going on. Brain affects mind and vice versa. Why should that be? It would seem that we are back to precisely the same mystery as was described earlier, the mystery of the effectiveness of mathematics, or of discovery versus invention. Both phenomena point to the same primordial unity of mind and matter, of mind and brain.

This ultimate unity of the mind and nature is a complex one. It is not a unity of identity where you maintain that the brain and mind are identical. Nor is it a simple duality. It is an ambiguity. The two realms of mind and body (or mind and nature) are self-consistent frameworks, each of which can be taken to be fundamental. Therefore, the attempt may be made to start with either one and reduce the other to it. Proceeding in this way consists of imposing an artificial unity by means of eliminating either mind or matter.

One way to think about this complex unity of mind and nature is by analogy to my discussion of the fundamental theorem of calculus.[15] Calculus is one unified subject that comes with two different contexts and a translation tool (the theorem), whereby changes in one domain are translated into the other. As I mentioned in chapter 5, the same phenomenon is the essence of the Taniyama-Shimura-Weil conjecture,[16] about which Barry Mazur commented:

> It is as if you know one language and this Rosetta stone[17] is going to give you an intense understanding of the other language. But the Taniyama-Shimura conjecture is a Rosetta stone with a certain magical power. The conjecture has the very pleasant property that simple intuitions in the modular world translate into very deep truth in the elliptic world, and conversely.[18]

A much more simplistic way of making the same point is by observing that in the gestalt picture on page 75, changing the young woman by giving her a necklace, say, changes the old lady, by giving her a mustache.

In each of the situations, neither interpretation is the "right" one. They are not identical but neither are they totally independent of one another. The two are tied together in a very subtle manner. The unity of mind and body we are talking about is upstream of both mind and body, and thus it is not necessary to reduce mind to body or vice versa. The mind/body problem is itself evidence for the existence of such a unified state, but unity is equally a resolution of the mind/body problem.

One last word about the three famous conundrums discussed in the last three sections—the unreasonable effectiveness of mathematics in the natural sciences, discovery versus invention in math and science, and the mind/body problem. Evidently they are not three distinct problems at all but variations on the same fundamental situation—the ambiguous unity that stands behind mind and matter, knowing and being, and subjectivity and objectivity.

UNITY AS THE CONTINUOUS AND THE DISCRETE

Science works with chunks and bits and pieces of things with
the continuity presumed, and [art] works with continuities of
things with the chunks and bits and pieces presumed.
—Robert Pirsig[19]

In science, we usually try to understand things by breaking them up into bits and pieces and then putting the pieces together again to form complex systems. We see the body as made up of cells, the cells as made up of molecules, and molecules as comprised of atoms. The realm of bits and pieces is the "discrete." In the preceding quotation, Robert Pirsig, the author of the best-selling book *Zen and the Art of Motorcycle Maintenance*, compares the realm of the discrete to that of the "continuous." He says that art can be distinguished from science—or in his terms, the romantic from the classical—because the former takes the continuous as primary, whereas the latter begins with the discrete. Though this may be true on one level, we cannot say that science ignores the continuous. On the contrary, the duality between the discrete and the continuous is the subject for a very old dialogue within science that continues to the present day.

Zeno's Paradoxes

The continuous and the discrete are two different and conflicting ways to think about unity. They each have enormous resonance and are basic to all human activity yet, as Pirsig intimates, they present radically different ways of understanding and experiencing the world. One could argue that the tension between the continuous and the discrete form the basis for the famous paradoxes of Zeno, like the one about Achilles and the tortoise. This paradox involved a race—say, a hundred-yard dash—between the speedy Achilles and a tortoise. The tortoise was allowed to start from the 50-yard line since Achilles runs much faster. Let's assume Achilles runs at a speed of 10 yards per second, and the tortoise is only capable of 1 yard per second. After 1 second, Achilles will be at the 10-yard line, the tortoise at the 51-yard line. Now 5 seconds after the start Achilles will catch up to the initial position of the tortoise, the 50-yard line. However, the tortoise will have moved ahead to the 55-yard line and so will be 5 yards ahead. It will take Achilles 0.5 seconds to get to this second position, at which time the tortoise will move slightly further, 55.5 yards from the start. And so things will continue like this. Each interval of time will see Achilles move up to the old position of the tortoise, while the tortoise moves just marginally further. Since an infinite number of these intervals exist, Zeno argued that Achilles could never pass the tortoise. On the other hand, Achilles is going so much faster it seems clear that he must win the race. (He finishes in 10 seconds, while the tortoise takes 50 seconds.)

This is one of the most famous paradoxes of all time. What is the point of the paradox and what is wrong with the reasoning that purports to produce two opposing conclusions? One of the questions raised by Zeno was the legitimacy of dividing time and space into an infinite number of segments—in fact, the legitimacy of dividing time and space at all! Zeno established these paradoxes in order to support the position of Parmenides, who "rejected pluralism and the reality of any kind of change: For him, all was one indivisible, unchanging reality and any appearances to the contrary were illusions to be dispelled by reason and revelation."[20] Zeno and Parmenides believed in unity as the most basic reality connected to the continuous, as opposed to the discrete. I shall expand on this point in the following paragraphs but, for the moment, let's just note that what we are calling unity has for millennia been connected to the question of the discrete and the continuous.

The word "continuous" can be defined as "forming an unbroken whole,"[21] and, of course, "whole" is another word for unity. In contrast, the noun "continuum" is defined as "a coherent whole characterized as a collection, sequence or progression of ... elements varying by minute degrees,"[22] which also highlights the idea of unity. Yet another dictionary has the following definition for continuity: "continuing without changing, stopping, or being interrupted in space or time" or "having no gaps, holes, or breaks."[23] It makes sense that wholeness or unity can have no holes or gaps, and as we shall see, this way of looking at continuity directly finds its way into mathematics and science. But let us step back from the definitions for a moment and ask what human experiences gave birth to the idea of the continuous. Where does this idea come from?

Sources of the Continuous

Time and space are the fundamental sources of the continuous. In science, they are fundamental; you can't define them in terms of anything more elementary. Time and space can be quantified through measurement and, in general, the continuous enters into science through the idea of number. In the case of the Greeks, numbers were, for the most part, geometric in nature, which tied them most intimately to the continuous.

But let's begin with space. We experience space as a vast container. It contains not only all the objects of the world but also all the processes. "Space is a container" is a metaphor, as Lakoff and Johnson[24] would say, that informs our experience of the natural world. It takes a particular form in the Newtonian idea that space is the three-dimensional Euclidean continuum in which every point can be uniquely located by an ordered trio of real numbers—its x, y, and z coordinates. Newtonian physics takes place within this continuous three-dimensional space, but the theory of relativity also takes place within a continuous space—in this case, a four-dimensional space-time continuum, which is also modeled on the same continuum of real numbers. These models are believable because we have a sense that the space that contains us and everything else is an object, a universal container. We seem to see that it extends in all directions; that it stretches out forever, seamlessly, with no holes or other interruptions. Space, we might imagine, is like the sky or a great ocean that contains everything. The objects of the world appear to exist within that space—we live and die in that space and all processes appear to happen within it. Whatever happens to individual objects or people, the space remains the same. It is not affected by the changes in the world. Space, conceived of in this way, is one root from which the continuous is derived. The metaphor of the universal container indicates that the sense of inclusive unity, the sense of being part of a larger entity, is at play in the idea of continuous space.

Time is another primordial source of the continuous. One of the most basic things about the world is change. Time arises from the experience of change but is not identical to it. Time involves the sense of flow. Events in the world appear to flow from the past to the present to the future, and time is a measure of this flow. But flow is a property that is usually associated with continuity. Since the age of Newton, time has been modeled on the continuum of the real numbers, just as space is. In the theory of relativity, time and space are unified and put together in one space/time continuum. But even before the advent of relativity, space and time were unified by the real number system.

Space and time are the basic sources of continuity. What they share is this property of unity and continuity. One way of making this intuition concrete is through using the real numbers as a common base for time and space. Of course, we must be careful to distinguish between the "continuum," the continuous as object and continuity, which is the continuous as process. So we might say that time is continuous and is represented by a one-dimensional continuum. Space is a three-dimensional continuum, but many people will also say that space is continuous. When you throw a ball in the air, you would describe its trajectory as continuous. This means it does not move from one position to another without going through a continuum of intermediate positions. In other words, the description of motion as continuous is usually done with reference to the two continua of time and space. Both are a manifestation of inclusive unity.

The continuous contains no gaps or interruptions because, like space, it seems to be whole, and a whole has no missing pieces. As we shall see, the property of having "no gaps" is a defining property of the real number system and the reason why the system of real numbers so effectively

models the continuous. This is another way in which the continuous refers back to the inclusive form of unity. When you say you are one with
somebody or some group, the essential thing is that, for the moment,
nothing separates you from the group. The focus is on the larger unit of
which you all form a part. Modern history is full of episodes of national-
ism and ideology, where individuals lose themselves in a country or race
or political party or merely in being a fan of some sports team. In these
cases, we can see how powerful the sense of inclusive unity can be.

Silence

If time and space are both sources of the continuous, another primal
experience could be mentioned as a possible, less obvious, source of the
continuous. This is the experience of silence. Silence is a container in the
sense that space is a container. When one gets out of the city and into
nature, one is struck, first of all, by the silence. This silence is not an
absence. On the contrary, it is a palpable presence—you can feel it. This
silence does not mean there is no sound; if a neighbor starts up their
chain saw, you can certainly hear it, but it seems as though the sound
emerges from the silence. Silence feels like a container. Silence has all of
the attributes that one associates with the continuous—in particular, it is
vast and unbroken. Silence is continuous but, in comparison, a particular
sound is discrete. It is interesting in this regard that when one returns to
the city and takes up one's busy life again, the silence that was so notice-
able in the country seems to disappear. It has not disappeared at all; you
are just not conscious of it. This is the reason that silence has for millennia been a way to evoke what I have been calling the blind spot, but it is
also a way to think of unity as a kind of universal context or background.
The conceptual world emerges from this background, but when you
focus on the conceptual world, the background is nowhere to be seen.

The Discrete

Now let me turn to the discrete. The objects of the world do not have the
same characteristics as space—they exist as well-defined entities *within*
space. The discrete arises as multiplicity, but multiplicity begins with
unity. Each object is a unity—it is one. But it is not the same one as the

one of space. When we perceive an object, we do so from the outside. We see it, hear it, or touch it. When we perceive space, we do so from the inside. Just as sounds appear to emerge from silence, so objects appear to emerge from space. They thrust themselves forward and demand your attention. The world does not present itself to us as undifferentiated chaos—it is made up of objects that form the primary datum of the discrete.

Usually, we think of collections of discrete elements as finite, and as such, we can count them. In this way, each finite discrete collection is represented by an integer. The set of counting numbers, {1,2,3,... }, although infinite, is usually thought of as discrete in mathematics. Nevertheless, for a collection to be discrete it must be made up of distinct individuals—that is, distinct units. The discrete involves the use of "exclusive unity."

The Real Numbers as a Model for the Continuous

I mentioned earlier that the statement that space is continuous is usually taken to mean it is modeled on real numbers. Recall (from chapter 2) that a real number can be represented by a decimal number, either finite as in 0.25 (1/4), or infinite as in 0.3333 ... (1/3), or 3.141592653589793 ... (π). Also recall that the usual picture of the real numbers consists of a horizontal line with each number occupying a unique location on that line (cf. figure 1, p. 26).

This picture is usually called the "real line," but the identification of the line with the numbers on it is metaphorical rather than literal. A problem arises though: Is a line continuous because it is parameterized by the real numbers, or are the real numbers continuous because they can be represented by a line? In fact, it is the identification, the metaphor, that makes the real line into a continuum.

Perhaps the line and the real numbers are both derived from a more primitive idea of continuity. Nevertheless, most people today think of real numbers and the geometric line as essentially identical. This means that the points on the line are identified with numbers. It is this kind of metaphorical thinking that enabled Descartes to build up his analytic description of Euclidean spaces where points are identified with numbers, curves and other geometric figures are seen as equations, and geometry (in general) is translated into algebra.

Real numbers have a privileged position in mathematics and science, especially the physical sciences. This is evident from the use of the word "real" to describe them. Why are these numbers more "real" than other kinds of numbers—the complex numbers, say? Why do we use the world "real" for these numbers as opposed to the more problematic words used for other kinds of numbers, such as "negative," "irrational," and "imaginary"? The real numbers seem to be a deep and far-reaching attempt to conceptualize an essential aspect of reality. That aspect is, of course, continuity—and beyond continuity, what I've been calling unity. Are real numbers merely one model of the continuous, or are they definitive? Do real numbers capture the essence of continuity? When we say that space and time are continuous, do we mean nothing more than the Newtonian idea that they can be described by the real number system?

The aspect of continuity that the real numbers capture is gaplessness, since the real number line has no holes. What does this mean? Suppose you divide all the real numbers into two sets **A** and **B**, with the property that every element of **A** is smaller than any element of **B**. You could imagine cutting a horizontal line with a knife, with **A** consisting of all points on the line to the left of the cut, and **B** consisting of those numbers to the right of the cut. For the real numbers, there is always exactly one number that corresponds to the cut, call it N. The number N is larger than every element of **A** (other than possibly itself) and smaller than every element of **B** (also possibly excluding itself). If there was a cut that didn't correspond to some specific number, we might say that the real numbers have a gap, but it turns out that the real numbers can be characterized by the property of containing no gaps in precisely this sense. On the other hand, the rational number line (the set of fractions arrayed on a horizontal line) have many gaps—for example, the square root of two corresponds to a cut where **A** consists of all fractions x such that $x^2 < 2$, and **B** of those fractions where $x^2 > 2$. Thus, the square root of two is a gap in the rational number line.

Today, some people challenge the privileged position of the real number system, but such challenges have a long and interesting history. The controversy about the nature and role of real numbers is more than a hundred years old. To understand it completely, you must go back a long way through the history of mathematics, but here I shall try to summarize the situation in just a few sentences.

The metaphor of the real numbers as a line, which I spoke of earlier, is very simple and self-evident. In fact, the identification of the real numbers with the picture of a line is almost too simple because it gives people the impression that the real number system itself is simple and easily understood. Yet real numbers are not simple at all—in fact, real numbers are one of the most complex creations of the human mind. Even today, all kinds of questions about real numbers are not understood, and remain unresolved. These are, for the most part, ordinary kinds of mathematical questions, such as the question about whether specific numbers are transcendental (are not the solution to any algebraic equation like $x^3 + 2x^2 + 3x + 1 = 0$).[25]

Real numbers have forced mathematicians and philosophers to confront other kinds of questions, too. These include the continuum hypothesis, which sounds straightforward but is irresolvable in principle. However, the basic problem can be seen in the following kind of paradoxical situation, which is due, in spirit at least, to Cantor. He showed that there are different orders of infinity. Whereas the counting numbers, the integers, the fractions, and the algebraic numbers are countable (can be put on an infinite list), the real numbers, the irrationals, and the transcendentals, are not. Suppose you say that a number is "knowable" if you can describe it in a finite way in words and/or symbols. Thus, every real number you can describe is "knowable": the root of two is knowable because it is the number whose square is two; pi is knowable because it is the ratio of the circumference of the circle to its diameter; and so on. However, it can easily be proved that "most" real numbers are "unknowable" in this way. In accepting the reality of real numbers, we have introduced into science entities that are intrinsically mysterious.

Gregory Chaitin[26] and others have even argued against continuity, and in favor of discreteness as a result of the various anomalies implicit in the real number system. His evidence comes not only from mathematics but also from physics and digital technology. In physics, there is the notion that physical quantities, like light, are quantized—that is, discrete. Of course, the digital computer is built on a finite universe of 0s and 1s. Thus, it is only capable of finite precision. For example, you cannot input an irrational number into a computer as an infinite decimal unless you do so in symbolic form, such as π or $\sqrt{2}$, or unless the data are somehow compressible, such as "the decimal expansion of the number consists of 1s followed by blocks of n 0s: 1,0,1,0,0,1,0,0,0,1,0,0,0,0, ... For "most" real numbers, this is not possible.

Given that most real numbers are non-computable, one might take one of the following two positions. The first is that reality is continuous, and that the discrete (including computer models) are approximations. The other is that reality is discrete and that the continuous is a different kind of approximation—what is called an interpolation. Interpolation is a word for the process of "filling in the gaps."

Many feel that the continuous in the form of real numbers is too basic to science to be reduced to the discrete even within theories of computation. In recent years, we have seen the development of a theory of computation that starts with real numbers. This theory was proposed by Lenore Blum, Felipe Cucker, Michael Shub, and Steve Smale in their book, *Complexity and Real Computation*.[27] They say, "The Turing model with its dependence on 0s and 1s is fundamentally inadequate for giving a foundation to the theory of modern scientific computation where most of the algorithms—with origins in Newton, Euler, Gauss, et al.—are *real number algorithms*." They maintain that their viewpoint is not new and trace it back to a 1948 statement by John von Neumann, who "was particularly critical of the limitations imposed on the theory of automata by its foundations in formal logic."[28] Von Neumann said that the problem with basing a theory of computation on formal logic was that "it deals with rigid, all-or-none concepts, and has very little contact with the continuous concept of the real or of the complex number, that is with mathematical analysis." Of course, the theory of computation proposed by Blum, Shub, and Smale is a mathematical generalization of the Turing theory, and contains it as a special case.

The important point is that this theory brings into question the assumption that the discrete is closer to reality than the continuous. As we have seen, the discrete and the continuous are both equally valid ways to approach the world, to approach unity. The computer has brought about a renaissance of the discrete because it seems closer to logic and "finite" modes of thought. But the synthetic and the continuous have been, and continue to be, a fundamental approach to science and mathematics, with their own contribution to make, and their own fundamental insight into the nature of things. The idea of building a theory of computation on the basis of the continuous seems to be a brilliant contribution, a radical reversal of the normal way of looking at things.

In summary, there remains today a continuing fruitful dialogue between the continuous and the discrete. The complexity of real numbers has given rise to doubts over whether the advantages of the real number system outweigh its drawbacks. In the work of Cantor, Gödel, and Chaitin, they have even forced us to contemplate the intrinsic limitations of human thought, which can be understood as the impossibility of conceptualizing unity in any definitive way. Real numbers bring with them uncertainty and incompleteness, as I have pointed out in previous chapters. Nevertheless, we cannot give them up because they are intrinsic to science and are our best entry into the continuous. Even if we could contemplate giving up real numbers, what could never be given up would be either the continuous or the discrete, for this distinction is built into reality itself.

Interestingly, two kinds of mathematics exist: the discrete and the continuous. Number theory, for example, is about a discrete object, the counting numbers, while differentiable manifolds—higher-dimensional generalizations of surfaces like the sphere and the doughnut—are continuous objects. Nevertheless, all subjects in mathematics use, to some extent or another, ideas and techniques that come from both the continuous and the discrete. It is often the interaction between these two that gives rise to fundamental insights. Physics too has room for the discrete and the continuous. Newtonian mechanics, for example, views reality as continuous, whereas quantum mechanics takes a discrete view.[29] It would seem that the debate about the discrete versus the continuous will never be resolved definitively—we have gained too much from both sides to ever give them up. So let's explore a different kind of resolution.

The discrete and the continuous are not opposites, even though one might think of them this way. In fact, they are ambiguous, although just saying this begs the question unless one plumbs the depths of what one means by ambiguity. An ambiguity is not just a duality. Furthermore, it is not enough to say that the discrete can be approached by means of the continuous, and vice versa—even though this is an interesting and fruitful position. An ambiguity must refer to a single concept or idea that can be expressed in two consistent but conflicting frames of reference. Here, the continuous and the discrete are the conflicting frames of reference. But what is the single idea that unifies them? It is reality itself. It is this elusive and complex unity that can never be definitively captured by thought.

The Origins of Unity in the Human Condition

When we discuss unity, we can see why science will never be understood by imagining that it stands outside of human life. On the contrary, what is basic and exciting about science is that science arises out of the human condition and reflects that condition in all of its complexity. Too often science positions itself as though the scientist were a pure observer, as though she were outside of the human condition looking in—the position that is the essence of what I called classical science. Such science recognizes one primary mode of being in the world—the observational mode. Yet when we study human beings, it is clearly the same human being who is both observer and observed. The pretense that only the observational mode is valid limits science in a fundamental way. I have been arguing that it is the discovery of these limits that characterizes the science of the last century and sets the challenge for the next. When scientists study human beings—when they study consciousness, for example—they obviously are not only on the outside looking in but also on the inside. It is equally true that this same inside–outside ambiguity has an effect on all of the scientist's interactions with the natural world. We affect the experiment by conducting the experiment. There is no privileged and totally objective perspective. Scientists interact with what they study. In certain circumstances, this interaction may be so subtle that it can be neglected for certain purposes, but it is nevertheless there. Since there is no such thing as a pure observer, the rules of the traditional scientific game break down.

The assumption that the scientist is an omnipotent observer is an understandable response on the part of the scientist to the existence of a fundamental human dilemma. The dilemma is that of self-consciousness, the dilemma of being simultaneously subject and object. This was discussed in chapter 6 as the ambiguity of participant versus observer, or in chapter 7 in terms of the existence of two cognitive modes. All human beings contain two legitimate ways of interacting with the world. Moreover, these two modalities are in conflict with one another.

Conflict

Conflict is, and always will be, an element of the human condition. In fact, it is conceivable that conflict is not just a characteristic of human life but pervades all of nature. In some significant areas of science, conflict already has a key role. It is a crucial element, for example, in modern evolutionary theory through the mechanism of the "survival of the fittest." It is also true that conflict, in the guise of competition, is seen to be a key element in economics. The Greek geometers had the brilliant insight that conflict in the form of contradiction could be used constructively as part of mathematical argumentation. But conflict cannot be restricted to particular disciplines; it underlies all disciplines. It is situated not only in the realm that is described by scientific theories, but also in the consciousness of the scientist who creates these theories.

At the existential level, you could put it this way: on the one hand, we are animals tied to the realities of the physical and biological worlds—in particular, we are mortal. On the other hand, we are the mind, a pure intelligence that seemingly stands outside of death and the natural world. As minds, we seem to transcend the world; we adopt a god-like stance of pure observer untainted by our own bodies. As anthropologist Ernest Becker[30] pointed out in his Pulitzer Prize–winning book, *The Denial of Death*, this duality and the conflict it produces is so painful that we habitually repress it. He even claims that such repression is the normal human reaction. But, in Becker's terms, this normal reaction is neurotic where neurosis involves the substitution of an artificial, and inevitably incomplete, construct for things as they are. This leads to his epigram, "To be normal is to be neurotic and to be neurotic is normal." That is, the normal way to deal with this conflict is to deny it, to pretend it does not exist. The same denial exists in science. It is the fiction of the pure observer, of the science of certainty. Classical science consists in the repression of the conflict that is inherent in the human condition. It does this by means of repressing the participant, the subjective.

Unfortunately, conflict cannot be willed away. Every human being carries within them a fundamental distress, not because they have done anything wrong, but merely because they are human. The alleviation of this distress is the basis of a great deal of religion and much of psychotherapy. Another way to (temporarily) alleviate this distress is through the creation of areas of consistency and coherence in our lives. The search for a way out of the conflict that characterizes the human condition can lead to the belief in a utopia where conflict has been vanquished once and for all, a utopia which may be the heaven of religion or a "theory of everything" that one finds in science.

You can think of science in one of two ways: either as an escape from the human condition or as the expression of that condition. This corresponds very well to the two approaches to science that I have been developing. It would seem to be inevitable for the human existential situation to find its way into science. What I have been calling the assumptions of clarity, certainty, and consistency in science constitute one way of responding to the human condition. In this book, I hope to bring the discussion back to where I believe it should be—science as intimately intertwined with the deepest aspects of the human condition and, in this way, an activity through which we can and do develop deep insights into the subtleties of who and what we are.

When the human need for certainty was earlier juxtaposed with the need for truth and illumination, a basic conflict was revealed. Recall Winer's statement about the human intolerance for uncertainty. We need stability and permanence; we need certainty, but we are presented with a world whose single rule seems to be transience, change, and impermanence. This is not a situation with which we are comfortable. A conflict exists inside each and every one of us. Our response to this conflict determines the nature of the science that we develop, and consequently the effect of that science upon society. Parenthetically, it is the fact that conflict is unavoidable to any realistic description of the natural world and the human condition that renders ambiguity (which necessarily contains a conflict between competing frames of reference) an appropriate tool for the task undertaken by this book.

Unity as the Denial of Conflict

Conflict is built into human life at its most basic level. Yet conflict is not the whole story. It must be set against another tendency, what we might call the intuition of unity. We have the intuition that there exists a state of harmony that we can discern both in ourselves and in the natural world. We now have these two primordial factors, intractable conflict, and perfect and absolute harmony that underlie science and human life. How can we possibly reconcile them?

We can see how the drama of unity versus conflict plays itself out in science by examining our own reactions to harmony and conflict. For the most part, we may be happy to acknowledge harmony as a goal or as an ultimate reality even if it sometimes seems like pie in the sky. Conflict is quite another matter. The statement that human beings are condemned to an existence that includes an unavoidable inner conflict is a very difficult pill to swallow. Something in us strongly rejects this idea. What is important is to investigate the nature of that rejection. How are we to deal with the ambiguity that pits unavoidable conflict against the intuition of harmony? One way is by denial, through imagining that the universe is spread out passively before our mind's eye like an intricate illustration or a movie. This stance essentially removes the conflict and, moreover, places the scientist in a god-like position, as though she were outside of the natural world. The benefit is a sense of being in complete control, a feeling of power. This is the origin of the disastrous alienation of humankind from the world, the feeling that we are the masters of nature and can mold it freely to our own purposes.

Yet what is often overlooked is that our rejection of ambiguity and its conflicts itself stems from an intuition of harmony and wholeness. The conflict between different cognitive modes masks a deeper conflict that arises because the very existence of two distinct modes of being contradicts our sense of ourselves as unitary, as one. It is not only that duality brings conflict in its wake, but that duality as such is unacceptable since it contradicts the intuition of unity. At the level of our conscious selves, what we are mostly aware of is conflict. Unity is most often hidden and when it does appear, it does so for a fleeting moment. We may look back on it as a peak moment in our lives or merely as a pleasant interlude.

If division and conflict are a fundamental condition of conscious human life then it would not be surprising that many of our activities are motivated by a desire to regain the unity that we have lost touch with but which is the ground of our experience and our deepest intuition. This is, according to the philosopher Søren Kierkegaard, the origin of the Old Testament myth of the Garden of Eden.[31] The Garden of Eden *is* unity, the ultimate mystery and the ultimate in simplicity. A writer can only put down various ways of speaking about unity in the hope that one of them will resonate with the reader's own experience. Ultimately, poetry and music are better in this regard than prose. Science is a grand adventure. It is the search for an all-encompassing unity. But the adventure is grounded in the unity that we wish to bring into consciousness. We seek unity because of our intuition that it is the deepest source of the self, the origin of who and what we are.

It is precisely because of this background of unity that we often respond by denying conflict, denying that duality is real. Thus, we insist on logical consistency. The demand for consistency is a natural reaction of our minds. Our brains have evolved from the need to see "the big picture." This could be seen as our response to unity; it is one aspect of our need for unity. We need to see things as consistent wholes because this is how unity appears to our rational minds. We then attempt to create a unity that is objective but artificial. We do this by filling in the gaps in our intellectual reconstruction of the world, and we do this because of our intuition that things are whole.

Then, we are shocked when someone, Gödel or Bohr, for example, discovers the gaps in our theories, the inevitable gaps in *all* theories. But our attachment to a particular theory was itself the consequence of the need for unity. Rationality and logic can be seen as an instance of unity. The drive toward the creation of "final theories" in science or of complete and consistent deductive systems is itself the manifestation of unity.

Ultimately, the a priori imposition of consistency does not work because it postulates a unity at an inappropriate level of objective rational theory. The unity we are all looking for lives behind or underneath the manifest and the objective. This is behind the metaphor of the blind spot and it is why I have needed to look beyond the classical in my description of science. Nevertheless, we can now see that the drive for unity as objective consistency lies behind a good part of the scientific enterprise.

We use the mechanism of denial on two levels. We do it to make ourselves feel better, to paper over our inner conflict. But the deeper reason is that we are in contact with a background of unity. Even denial and repression come from unity. In both of these ways, unity is the root cause of our scientific investigations. We are pushed ahead by our need to resolve our own inner conflict. But we are simultaneously pulled along by our sense of a fundamental unity that lurks seemingly just beyond our reach. Robert Winer expresses this situation in a psychotherapeutic context: "Both patient and therapist will be swept along in this venture by the human longings for integration, continuity, and narrative closure and the human intolerance for uncertainty."[32] Whether pushed by uncertainty or pulled by the need for integration, either way the prime factor is unity.

Spinoza spoke of unity as "the knowledge of the union which the mind has with the whole of nature." Jill Bolte Taylor said that, "the boundaries of my earthly body dissolved and I melted into the universe."[33] Of course, Taylor's condition comes as a result of losing contact with the analytic capabilities of her left hemisphere. Compare this with Alan Lightman's description of the moment of creative breakthrough:

> I woke up about five a.m. and couldn't sleep. I felt terribly excited. Something strange was happening in my mind. I was thinking about my research problem, and I was seeing deeply into it. I was seeing it in ways I never had before. The physical sensation was that my head was lifting off my shoulders. I felt weightless. And I had absolutely no sense of my self.… I had no sense of my body. I didn't know who I was or where I was. I was simply spirit, in a state of pure exhilaration.[34]

The sense of unity in both cases involves an altered sense of self—a loss of some more restricted view of who we are. However, this loss is not merely the loss of the analytic capabilities, otherwise what was Lightman doing at his desk? The point that is made quite well by these stories is that unity is not something outside of yourself that you look at or describe objectively.

The science of wonder is a science that is saturated with the vision of unity. In this way, it is true both to its own subject matter but also to the reality of its own creators. It does not create an alternate universe that replaces the contingency of life with the certainty of the absolute but stays true to the ambiguity of the human condition, the ambiguity of consciousness itself, the ambiguity of unity.

Creativity as the Emergence of Unity

The interesting thing about moments of creativity is that they involve an experience of unity—the subject that you are studying or the problem that you are trying to solve suddenly fits together in a way that was previously unexpected. A new idea or a new approach appears, seemingly out of nowhere. It supplies a new perspective. New connections are revealed that were not previously suspected. So the objective situation is transformed. Often, the power of the new idea is such that you can immediately imagine ways of applying it to other situations, of abstracting the idea and generalizing it.

However, the unity that appears does not only live in the obvious objective situation and this is where Lightman's testimony (or the famous account by Henri Poincaré)[35] is so valuable. These creative moments are accompanied by a sense of assurance, of knowing that there is something of value going on. The sense of unity that emerges arises in both objective and subjective domains. Unfortunately, only one of these domains is usually considered to be a part of science proper, yet the creative experience itself might be considered as the expression of unity. A science of wonder has a place for the creative process; it has a place for unity. But it is a unity that is dynamic rather than static—it changes and grows, it is a unity in continual evolution. The reason for this dynamism is precisely that unity is continually divided against itself and then reunited in new and unsuspected ways in an ongoing symphony of creativity.

CONCLUSION

This entire discussion of unity, Oneness, the discrete and the continuous can be summarized very succinctly by a famous Japanese haiku poem:

> The old pond,
> A frog jumps in,
> Plop![36]

Here we have the inclusive and the exclusive, the continuous and the discrete. But why talk about it? The poem is enough.

9

~

The Still Point

At the still point of the turning world. Neither flesh nor fleshless;
Neither from nor towards; at the still point, there the dance is,
But neither arrest nor movement. And do not call it fixity,
Where past and future are gathered. Neither movement from
   nor towards,
Neither ascent nor decline. Except for the point, the still point,
There would be no dance and there is only the dance.
—T. S. Eliot, "Burnt Norton"[1]

INTRODUCTION

This chapter is an attempt to integrate some of the themes that have arisen in the previous chapters and, as a result, to begin thinking about science in a different way. A new perspective on science should have room for both the objective and the subjective, for certainty and reason, but also for the ambiguous and the paradoxical. It should view science as dynamic and creative rather than objective and unchanging.

Any attempt to describe the world precipitates, implicitly or explicitly, a spiral of self-reference. This chapter will show how self-reference and the blind spot are implicit in ambiguity, and this will lead to a new fractal perspective on science. Ambiguity was the theme of chapters 5 and 6, but the discussion is still not sufficiently subtle for my purposes since a basic problem arises from a serious consideration of ambiguity. It is simply this: If you now feel that you "understand" ambiguity, then the situation has lost its ambiguity for you. One essential ingredient in ambiguity is the incompatibility between the various perspectives, and it is precisely this sense of incompatibility that is lost when the ambiguity is resolved and the ambiguity is resolved when the situation is understood. Understanding *is* the resolution of ambiguity. If we lose touch with the conflict by "understanding" ambiguity, we shall have lost an essential aspect of the ambiguity.

Ambiguous situations have two poles, complementarity and conflict. Ambiguities are dynamic because they tend to move from conflict to complementarity. Nevertheless, we normally focus on the resolution and not the situation that evoked the resolution. We like our ambiguities to be resolved.

In other words, not only does ambiguity refer to situations with an incompatibility, but ambiguity is itself a situation that contains such an incompatibility because the two aspects of conflict and complementarity are themselves in conflict. If we were to understand ambiguity, then the incompatibility would disappear and we would be describing something else—flexibility, perhaps, or the existence of multiple perspectives—but not ambiguity. The most basic thing about ambiguity in the way I am using the term is that it is "generative" not static; the ambiguity generates potential resolutions. Any specific definition or particular understanding of ambiguity must necessarily be tentative. That is not to say that you cannot get a feeling for ambiguity—of course, you can, and hopefully that has happened to all careful readers of the book. But you cannot pin down that feeling in an airtight and all-inclusive definition. In other words, ambiguity is real but cannot be made precise. This is what I said characterized the blind spot—something that is real but cannot be grasped, and in this way ambiguity evokes the blind spot and is similar in this regards to the ideas of zero, infinity, and randomness.

It is ambiguity and not certainty that best describes the way things are. Of course, any situation that is ambiguous contains an element of uncertainty, but the *fundamental ambiguity* that I shall describe next also contains an irreducible element of self-reference. This will demonstrate the interdependence of these fundamental conditions—uncertainty, self-reference, ambiguity, and the blind spot. In the process, it will erase any clear line of demarcation between the objective and the subjective. The discussion of science will thus be pushed toward the eye of the storm that stands behind any serious attempt to describe the situation of the scientist who attempts to get correct inferences about the world within which she acts as a conscious participant.

Not only does the world of science contain this self-referential spiral, but precisely the same difficulty lies in wait for the reader who makes an authentic attempt to plumb the depths of ambiguity—and in this chapter I am pushing toward those murky depths. Now any discussion of ambiguity puts a large burden on the reader because ambiguity is not a concept that can be understood in a conventional manner without eliminating the tension and dynamism that is present in ambiguity. Ambiguity is ambiguous, and this element, which may be seen by the rational mind as paradoxical, is pushed to an extreme in the following discussion of the "fundamental ambiguity." The intent is not only to present a theoretical position but also to induce in the reader an internal state of ambiguity out of which a new view of the world and of science may hopefully arise.

THE PROCESS OF BECOMING

The emergence of uncertainty as a factor that cannot be removed from even a scientific description of the world comes as a result of a change of emphasis in science from "being to becoming" as the Nobel Prize–winning chemist, Ilya Prigogine, has said. One might think of this as a change in emphasis from structure to process. Process is inseparable from the view of things that this book is presenting.

Static models in which things are in some sort of equilibrium are very attractive. At one time, there were many people who believed in a steady-state model for the universe, a state of equilibrium where comings and goings sort of balanced out. Of course, this model has lost out to the big bang theory. An even more fundamental change in scientific thinking arrived with the theory of evolution, which presented a totally new picture of how the natural world functions, and of the position of humanity within that world. The paradigm shift here was from a static view to one that is dynamic and emphasizes change. In much of science, classical theories tended to focus on states of "equilibrium" and not so much on what led up to the equilibrium state or on how that equilibrium might break down. In recent years, however, with the development of the dynamical systems approach in subjects that go under the names of the "theory of chaos" or "complexity theory," we have a switch in emphasis to states that are not in equilibrium. There is, for example, Prigogine's work on dissipative structures and their role in thermodynamic systems that are far from equilibrium. Taken together, all of these theories represent a huge shift in emphasis in modern science.

The debate between being and becoming is not a new one. Is the world primarily made up of "things" such as quarks, atoms, molecules, or cells? The normal way to think about the natural world is in terms of a hierarchy of forms, the "atomic" way of picturing the world that goes back to the Greek thinkers like Leucippus, Democritus, and Epicurus. The search for the most elementary substance (today they might be "strings") is still the dominant tendency in modern science. However, there has always been another tendency, one that says change is more fundamental than the things that change. This point of view takes process to be its most basic object of study. Objects are then seen as stable configurations within that process. It is like looking at a turbulent stream. States that are relatively stable, like whirlpools, arise due to the rate of flow of the water and other constraints on the system. These whirlpools, or other configurations, are relatively stable, but are not considered to be fundamental since they can disappear if there is a variation in the rate of flow. What is primary in this view is the process, the flow itself.

The latter point of view, which goes under the name of "process philosophy," begins with the Greek philosopher, Heraclitus of Ephesus (b. ca. 540 BC), who famously said that you could not step into the same river twice.[2] Thinkers whose writing resonates with such a process point of view include Gottfried Leibniz, Henri Bergson, Charles Sanders Peirce, William James, and A. N. Whitehead. This book also contains a strong echo of such a position, since a process orientation tends to emphasize such things as contingency and creativity.

One question that arises in this context is "If process is the fundamental observed characteristic of the natural world, then what is the relationship between formal, and inevitably static, scientific theories and the world of change they describe?" How, for example, does the theory of evolution, a theory that is usually regarded as established, describe actual, ongoing evolution? The strange thing is that the change being described disappears at the theoretical level; we describe change using an unchanging theory of change.

By discussing things in terms of words and concepts, equations and symbols, we necessarily freeze them. They are like parameters in mathematical expressions—they remain constant (by assumption) for the duration of the discussion. This is the deeper meaning of the logical "Law of Identity, $A = A$." If things are static, this law is obvious. If "A" is continually changing, then it is not so obvious. Do we know that each occurrence of "A" is referring to the same unchanging substance? If each occurrence of "A" were different from every other, then not only would every electron differ from every other but even a single electron would be different from one moment in time to another. In such a situation, science and mathematics would be impossible. To talk reasonably about things, we need to freeze them for a certain period of time at least. One might then say that we conventionally operate "as-though" this frozen world of concepts and objects were permanent. The scientific world of particles, atoms, and molecules would then be based on this "as-though" condition, on the assumption that this "freezing" is a valid way to understand the world.

For the moment, let us just note that there are these two possible ways of looking at the world—a world of objects versus a world of change. These two points of view each have great resonance within modern science. Of course, there can be intermediate views that attempt to reconcile the dichotomy that has been described by establishing a kind of "complementarity" between these two views. However, it is in the spirit of this book to consider such dualities as potentially generative ambiguities. Therefore, I shall not jump to the conclusion that one of these views must be right and the other wrong, or to some attempt to eliminate their conflicting claims to priority. Both views are interesting; both are valuable; both have something to add to our understanding. Yet there is an irreducible and irreconcilable conflict here. This chapter will attempt to do justice to both of these positions and to develop a way of thinking that has a place for them both.

Dynamism without Movement

Is there anything that stands prior to both elements of the preceding dichotomy? This is where the quotation with which the chapter began comes in. When the Nobel Prize–winning poet T.S. Eliot refers to the "still point," he is evoking "neither arrest nor movement," but rather a sort of pure dynamism that is upstream, so to speak, of both change and stasis. Now, for Eliot, the still point is not a state among other states. It is a revelation—it is primordial, for without it "there would be no dance and there is only the dance." It is at the origin of change, and change is all that there is. Without it, there is no phenomenal world, nor is there a mental world. The still point is pure potential, the potential for all specific things and specific thoughts. It is not static. In fact, you cannot even "know" it, because knowing it would make it into something stable, some object or concept, and this would contradict its essential nature, which is dynamism, without form or movement. Elliot's "still point" is a vision of a reality that is pulsating with energy, but is under perfect control. It is stasis under stress: the irresistible force encountering the immovable object, thereby generating enormous tension.

From another tradition entirely, there comes a quotation from Dogen (1200–1253), the seminal figure in the cultural history of Japan:

> Being-time means that time is being. Every existent thing is time. You must recognize that every thing, every being in this entire world is time.… Do not regard time as merely flying away; do not think that flying away is its sole function. Because you imagine that time only passes, you do not learn the truth of being-time.

In this difficult passage, he explains his insight into what he calls "being-time"—a dynamic force that is the essence of both objects and process, through which the world is organized. Dogen warns us, "do not regard time as merely flying away"—that is, this dynamism exists now not in the movement of some substance that we habitually call time. As far as I can make out, this is exactly what Eliot intends to convey by the expression "the still point." He makes this explicit in the following lines of the poem when he says, "And I cannot say how long, for that is to place it in time." He means that our normal conception of time, time as minutes and hours, as days, weeks, and months, arises subsequent to something more basic—a pure dynamism—that he calls the still point and Dogen calls being-time.

This dynamism is related to what Oliver Sacks had to say in chapter 7, "the world is organized … by the power of number." This amounts to seeing "number" as a dynamic organizing principle rather than a passive object. The proto-concepts of science—space, time, number, energy, and so on—are all dynamic forces, each sharing the essence of Elliot's "still point." When we think of these things as concepts, we identify them with some explicit definition. This has the advantage of stabilizing them and making them static (until such a time as the definition is changed, of course). When we think of them as proto-concepts, as the generators of concepts, they are dynamic. This chapter (and the whole book, really) is about that ungraspable dynamism that generates the scientific world, but it is also about the relationship between that dynamic state and the more or less static concepts that form the basis of our theories. The last chapter was about unity. The essential aspect of unity is that it is not a stable object or state that one can point to. Unity is dynamic. This dynamism is one thing that I want to investigate in this chapter.

However, it is not just these foundational elements of science that are dynamic in nature, science as a whole is dynamic. Science is something you do! It is alive, vibrant, and sometimes even revolutionary! In the last five hundred years, science has utterly transformed the earth, changed the nature of human societies, and radically altered the way we think about human nature and the nature of life. Science is inseparable from change, not a safe haven that can protect us from the dangers and challenges of change. Science is not merely a picture of reality, something that you read about in comfort. It is not only a list of universal laws; it is not even a blueprint for the universe. The nature of real science, in short, is change not stasis. Yet when we think about the nature of science and when we apply science to technological projects, we tend to think of science as something fixed, objective, and absolute, a way to control the human environment, a way of removing the contingency that is inherent in life. There is even a fairly common belief that science and technology are the only means with which to save ourselves from the intractable problems we face as individuals and as a global society. At its most extreme, there is the hope that science may one day save us from the ultimate and most terrifying difficulty of all—death.[3] Many of us, in one way or another, hope for some sort of immortality through science. But this is a simplistic notion of immortality, and is intimately connected with a science that produces a static and unchanging truth. Elliot, Dogen, and many others are pointing to a different kind of immortality—the immortality of the moment, the immortality from which time arises—in other words, the still point.

These contrasting pictures—of a dynamic science versus one that is essentially static—is a way of restating the two views of science that I have been developing in this book: the science of wonder versus the science of certainty. The science of wonder is dynamic, and the science of certainty is essentially static, but the view of science that is being developed here is of a dynamism that contains both stasis and change. The challenge is to find a way of thinking about science that is rooted in the process of doing science, and therefore the reality of a universe of continual change, a physical and biological universe that is never still, that knows no other state other than that of continual moment-by-moment evolution.

Science and Change

The key to approaching this seemingly impenetrable barrier that divides what is real from the description of what is real, the key to reconciling the essentially static nature of scientific theory with the dynamic nature of the real world, lies in the elusive notion of ambiguity. I say "approaching" the barrier and not "crossing" it because, as I have discussed at length, reality is ultimately ungraspable. Nevertheless, the attempt to grasp the ungraspable is the quintessentially human act. It is also the means through which we transcend the human condition by losing ourselves in something that is much vaster than our individual lives.

Ambiguity is, in a certain sense, the ultimate attempt to grasp the ungraspable. Describing ambiguity, as I said earlier, is itself paradoxical. To the extent that it succeeds, it fails; to the extent that it fails, it succeeds. In what follows, I shall try to work with this paradoxical situation and examine what, if anything, will come out of that study. My goal is ambitious, nothing less than to formulate a new paradoxical, but fecund, basis for considering the nature of scientific activity—a new kind of philosophy of science. This philosophy will emerge from the still point, from pure dynamism or dynamic unity. In fact, pure ambiguity is the closest we can come to describing the still point. I shall do this through a final reformulation of the ambiguity of ambiguity, which will be seen to contain both movement and stasis, both ambiguity and clarity.

Normally, we imagine that the world is unambiguous—that the true state of things involves a condition of non-ambiguity. We believe that this state exists independent of our knowledge or our lack of knowledge of it. You might call this belief the first axiom of science, the "Axiom of Clarity." Without assuming the validity of this axiom, most people would be hard-pressed to say what it means for a theory or activity to be "scientific." And yet we have seen that a good deal of modern science contains elements of ambiguity and self-reference. As the frontiers of science move toward the biological and cognitive sciences, this tendency toward the kind of self-referential ambiguous situations that I have been discussing cannot fail to be augmented. Thus, there is a discrepancy between the story that science tells about itself, which is monolithically unambiguous, and the realities of life on the ground, so to speak, in which science is intrinsically self-referential, creative, and ambiguous.

What I am going to do is reverse the situation between ambiguity and clarity. As I mentioned earlier, I shall consider the possibility that fundamentally things are ambiguous and that the unambiguous is merely one particular point of view that arises out of a more basic ambiguity. My jumping off point will be a formulation from Albert Low (shown in the following excerpt).[4] Low's statement is perhaps as close as one can come to formulating the mysterious relationship between what can and what cannot be known. For those readers who would like to pursue further the relationship of ambiguity to evolution, creativity, and consciousness, I recommend the books by Low listed in the references.

The Fundamental Ambiguity
Fundamentally there is an ambiguity, one face of which says that there is an ambiguity while the other says that there is no ambiguity, but this face itself is not unambiguous.

Notice right off that the fundamental situation of ambiguity that is being described, a situation containing both ambiguity and clarity, runs parallel to the earlier discussion of a dynamism that contained both stasis and change. In fact, there is a deep connection between dynamism and ambiguity that will be brought out in what follows.

I first read Low's statement of the fundamental ambiguity a number of years ago and was immediately intrigued by it, in part because of the element of self-reference—ambiguity within ambiguity—that it contains. As we began to see in chapter 6, the element of self-reference has the potential for generating great complexity. So it is with ambiguity, since—in this formulation, and as we have seen in other parts of the book—ambiguity is ambiguous. The statement also says that at the most fundamental level, clarity is a particular aspect of ambiguity. This is most peculiar. So let us begin by looking closely at the statement of the "fundamental ambiguity" and attempting to draw out its general meaning and its complexity. Then, I shall look at the implications of this statement for our understanding of science.

What strikes you first about the statement is that it reverses the usual roles of ambiguity and non-ambiguity. Normally, we think that the world is unambiguous and that ambiguity arises either from ignorance or from error. The statement of the fundamental ambiguity says that, on the contrary, ambiguity is fundamental and that non-ambiguity is merely one aspect of this fundamental ambiguity. You could just say that the statement is nonsense (since it definitely has a paradoxical flavor) and leave it at that. Doing this, as we shall see, would actually be in partial agreement with the statement itself. Of course, I will not stop here since, even though the statement is initially obscure, it contains some startling implications.

The second thing we might notice is that what the statement calls the two "faces" of ambiguity and non-ambiguity are in complete conflict with one another. One says there is no ambiguity, whereas the other asserts there is. Thus, the statement is in accord with our characterization of ambiguity as involving two conflicting frames of reference.[5]

But let's begin with the statement as a sentence. Is it clear or is it ambiguous? At this level, you must say that it is clear. It has a precise meaning in terms of the English language; it is well formulated; the logic is straightforward. This is the face of the statement that is unambiguous, the face that represents a kind of theoretical hypothesis that could be tested against empirical standards. However, if its objective meaning is not ambiguous, then it contradicts itself when it says that only one face says there is no ambiguity. In this way, the clarity of the statement is not unambiguous. In this regard, it is reminiscent of the "Liar's Paradox" attributed to Epimenides which says of itself that, "This sentence is false."

Now look at the statement again and, this time, ask, "What does it mean?" When applied to itself it says that its own meaning is ambiguous (i.e., any attempt to understand it is a priori doomed to failure). It appears to have a built-in self-destruct mechanism. Remember that this was possibly one's initial response to this statement or, really, to any statement that appears obscure on first reading.

But this statement seems to say about itself that it *cannot* be understood.[6] (In this way, it follows along the central argument that Kurt Gödel used in his famous incompleteness theorem.) Thus, the meaning of the statement is obscure, difficult to grasp conceptually, and, in a word, ambiguous in the popular sense of the term. At this level, the face that says there is no ambiguity is denied.

The statement is ambiguous in one sense and unambiguous in another. This is, in fact, consistent with the statement. Thus, we are back to thinking that the statement is generally true and clear, that the face that says that there is no ambiguity is the right one. This puts us back where we began, in an endless spiral of ambiguity and clarity. Therefore (at a deeper level this time), the statement is ambiguous since we cannot determine whether it itself is true or false, clear or unclear. The statement is an accurate evocation of the situation that it describes. It itself contains this ambiguity within ambiguity.

Now let us reflect back on the previous discussion. Notice that the attempt to deal with the original statement in the way I have described is a process. First, we look at it grammatically and discover that at that level it makes sense. Then, we look at the meaning and it does not. Then, we look at the statement again in the light of what we have just done, and now it makes sense again. But this very coherence is at variance with the statement. And on and on we go, deeper and deeper into ambiguity and clarity. Is ambiguity ambiguous or is ambiguity unambiguous? Can this situation be resolved? Well, if it can, then the resolution would make the situation unambiguous and therefore deny the truth of the statement. Right away, one gets a glimpse of how this formulation of ambiguity is dynamic and not static—you just cannot pin it down!

Notice also that the process of plumbing the depths of the statement is neither really objective nor subjective. The statement may originally seem to be objective but even when we ask what it means we are introducing an element of subjectivity. Moreover, in subsequent steps, we are always comparing the statement to our understanding of the statement we obtained by going through the previous steps in the process. We are implicated at every step in the process and our involvement becomes part of the subsequent process. This process we are describing is one that

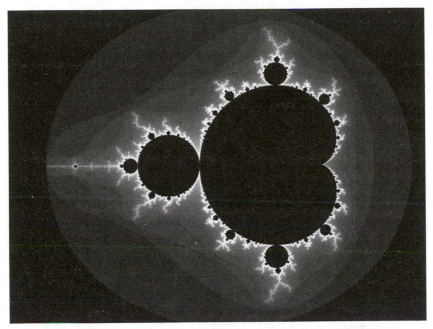

Figure 3

goes on in us, in our minds. But we cannot say it is nonobjective. After all, there is the statement. If you think about it, the spiral of objectivity and subjectivity reflects the spiral of ambiguity and non-ambiguity. Could they be two ways of referring to the same basic situation?

Metaphorical Pictures of the Fundamental Ambiguity

Let us return to the description of the "fundamental ambiguity" once more and develop a couple of ways to picture the situation it describes. These descriptions will be geometric. The first is a fractal and the second is a kind of description that comes from the theory of chaos. Now, a fractal is one of those beautiful, infinitely complex, and seemingly mysterious mathematical objects that are self-similar, which means that every tiny piece of the diagram is a scaled-down version of the whole. In a way, every fractal is a visual representation of the infinite complexity that is implicit in situations of self-reference. The most famous fractal is the Mandelbrot set (see figure 3). Fractals, although they look extremely complex, are often generated by simple mathematical formulas that are

Figure 4

applied iteratively, in a form of self-reference that involves repeating the same procedure or sequence of procedures over and over again (where the calculation at a given step depends on the values that were calculated in the preceding step or steps).

The first geometric model of the fundamental ambiguity is a variation on the famous yin–yang symbol of Chinese Taoism (figure 4). In this diagram, ambiguity is coded in black and non-ambiguity in grey. The outer circle, the circumference, which is to be identified with the entire ambiguous situation, is black. The interior of the circle divides into two smaller figures that represent the two faces of ambiguity and non-ambiguity. However, each half contains a figure of the opposite color, a circle, within which the process of subdivision begins again. This generates a cascade of grey and black circles where all of the black circles are essentially identical (except for a scale factor) to the original circle. When you say that the picture is a fractal, you are referring first of all to the fact that the cascade of circles within circles goes on forever, and to the fact that the overall situation has a kind of symmetry that I referred to earlier as self-similarity—the inner structure of any circle is essentially identical (except for its initial color).

Now the yin–yang diagram does not perfectly capture the situation described by the fundamental ambiguity. The problem is that the picture has an apparent symmetry between grey and black regions, whereas in the statement there is no symmetry between the ambiguous and the unambiguous. Thus, the diagram may seem to imply that the ambiguity and non-ambiguity are complementary, when in fact they are in conflict. Nevertheless, it is a useful diagram because we can clearly see in it the fractal nature of the situation. It is interesting in this regard that Niels Bohr, the main author of the standard model of quantum mechanics, had the yin–yang symbol on his family coat-of-arms. For him, it captured something essential about the view of the world that comes out of a deep consideration of the implications of quantum mechanics, and in particular, of the property that is called complementarity. Now the usual yin–yang symbol is not the same as the one in figure 4, since it stops with the smaller white and black circles within the curved black and white regions. Ours never stops. Nevertheless, it is possible that the fractal yin–yang diagram would be a better representation for the world of quantum mechanics. This is also implied by the discussion in chapter 5.

Figure 5 is another way of representing the fundamental ambiguity, one in which the fractal nature of the situation is implicit not explicit, as in the previous diagram. The diagram is an example of what is called a "linear graph," which consists of a number of nodes (dots) and various arrows that go from one node to another (or from a node to itself). A *path* in a linear graph consists of a sequence of nodes that have arrows between them written in an order that is consistent with the direction of the arrows. In our graph, IJJJIJ is a path, but JIIJ is not (in our graph, every occurrence of the node I *must* be followed by the node J). A *loop* is a path that starts and ends at the same node—for example, IJJJIJI. J itself is a loop, as is JJJ, IJ, and JI.

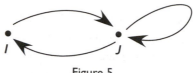

Figure 5

In figure 5, nodes I and J of the graph stand for the two faces of the fundamental ambiguity. "I" represents the face that is unambiguous, and "J" the face that is ambiguous. The arrows stand for something like "contains," or even "is" when it is used metaphorically. For example, the statement "clarity is not unambiguous" translates into an arrow from I to J. The loop from J to J says that ambiguity is ambiguous, whereas the arrow from J to I says that the unambiguous is one face of the ambiguous. Thus, the diagram captures a good deal of the complexity of the situation. In particular, it captures the non-symmetry between ambiguity and clarity.

Figure 5 is well known from the theory of chaotic systems and is associated with situations of great complexity whose behavior is called "chaotic."[7] The complexity of the situation is generated by the fact that the graph contains various paths that can be repeated as many times as we please. Two of these "loops" exist, the one from J to itself, the other from J to I and back to J. You can think of these as iterations, as feedback loops, or as a kind of self-reference. We have these feedback loops of arbitrary length since IJJJ … JI is always a loop, no matter how many J's we have in succession. The advantage of this picture is that it shows why ambiguity is a dynamic principle. Many ambiguous situations harbor the potential of generating just such an infinite process. In fact, this infinite cascade associated with the fundamental ambiguity is similar to the infinite cascade of observing yourself as observer or participant that I introduced in chapter 6. By the way, both of these examples show how infinity, in the guise of such feedback loops, is an intrinsic part of the human condition, and not just an artifact of scientific mathematical language.

In the rest of this chapter, I shall consider the various aspects of the dynamic situation that we are calling the fundamental ambiguity. Something about ambiguity always slips from our grasp, making it ungraspable. This is the aspect of ambiguity that is ambiguous. Yet there is also something we can understand. When I discussed the way in which multiplication is ambiguous, we could understand the ambiguity. We could describe it as a kind of flexibility. So already in my discussion of ambiguity I have touched on these two aspects of ambiguity—what I am now calling clarity, and what I previously called ambiguity.

Let me go on and, in line with our usual procedure of beginning the discussion with the "normal" point of view and not with the position that is logically the most basic, let me take up the unambiguous. Let us think about clarity.

Ambiguity and Clarity

The world is out there and what is out there is clear and unambiguous.

This is how things appear to us. It appears to be an obvious statement of fact. Of course, we recognize that two paintings of the same scene may well differ, and even two photographs will be different depending on the quality of the light and the viewpoint from which they are taken. Nevertheless, we believe there exists a unique objective view of things—essentially, the way things are. The certainty that we strive for in science is based on the existence of this domain of clarity.

In contrast with the physical world, our mental universe is normally anything but clear and distinct. Consciousness has often been likened to a "stream," and those who have any experience in introspection might say that one's interior mental space is more like a chaotic torrent than a sedately flowing stream. Mental constructs like words, thoughts, feelings, sounds, and visual images seemingly arise out of nowhere, evoke other mental objects, and disappear. Calling these mental constructs objects is already giving them a solidity that is not consistent with the experience. If we choose to use the metaphor of a flow, then mental objects could be called crystallizations within the flow. They are not necessarily clear. It takes language and a lot of work to make aspects of this inner world solid and clear enough to communicate to others—even to make it appear coherent to oneself. Doing science, especially theoretical science, requires the creation of stable coherent mental structures that model the coherent picture that our senses and our minds create out of the natural world.

At first glance, the external world appears to be coherent and clear, whereas the mental world is not. A closer inspection, however, demonstrates that things are not so simple. The domains of clarity that theoretical science builds up to correspond with the supposed clarity of the natural world are constructed through the use of language, mathematics, and logical reasoning. Mathematics itself is a paradigm of a domain of clarity and the model for other disciplines. However, the clarity of mathematics applies only to its formal surface structure, with its logic and precision. That even formal mathematics contains ambiguity was a point I made in chapter 5. The claim that ambiguity exists within mathematics and science is not really that controversial. What is controversial is that there may exist irreconcilable conflicts between various points of view, that it may not be possible to contain all possible frames of reference within one unified logical superstructure. If such a mega-structure exists, even if it only potentially exists, we are in the science of certainty. Clarity may only be local not global. As David Bohm said, "all theories are insights which are neither true nor false but, rather, clear in certain domains and unclear when extended beyond these domains."[8] Ambiguity arises on the common boundary between two theoretical models, like the one in physics between the theories of relativity and quantum mechanics.

Science appears to be a domain of clarity at first, but closer examination shows us that the clarity of the mathematical and scientific domains contains ambiguity. The view from the perspective of the researcher is that one is standing on a small island of clarity in the midst of an ocean of obscurity. This is in correspondence with the statement of the fundamental ambiguity, "fundamentally things are ambiguous."

It is certainly conceivable that the clarity we perceive in the world is something we bring to the world, not something that is there independent of us. The clarity of the natural world is a metaphysical belief that we unconsciously impose on the situation. We consider it to be obvious that the natural world is something exterior of us and independent of our thoughts and sense impressions; we believe in a mind-independent reality. Paradoxically, we do not recognize that the belief in a mind-independent reality is itself mind-dependent. Logically, we cannot work our way free of the bubble we live in, which consists of all of our sense impression and thoughts. The pristine world of clarity, the natural world independent of the observer, is merely a hypothesis that cannot, in principle, ever be verified.

To say the natural world is ambiguous is to highlight this assumption. It is to emphasize that the feeling that there is a natural world "out there" that is the same for all people at all times, is an assumption that is not self-evident. This is not to embrace a kind of solipsism and to deny the reality of the world. It is to emphasize that the natural world is intimately intertwined with the world of the mind. In consequence, the natural world is a flow just like the inner world. By stabilizing the inner world through language, logic, mathematics, and science, we simultaneously stabilize the outer world. The result of all this is the recognition that the clarity we assume to be a basic feature of the natural world merely masks a deeper ambiguity.

One of the functions of mathematics and science is precisely to deny this ambiguity. This is really the motivation behind the science of certainty. It occurs when one attempts to give a description of the "scientific world" as objective in an absolute sense without trying to see "objectivity" as one way in which the world appears to us because it is one particular mode of consciousness. Subjectivity is, by and large, not addressed by science except by "objectifying it" by, for example, showing how it arises out of a certain activity of the brain.[9] This has the effect of eliminating the ambiguity that I was discussing, but objectifying subjectivity in this way does not produce subjectivity in the same way that saying happiness involves the release of certain chemicals in the brain does not tell you what happiness really is. It is much more likely that objectivity and subjectivity cannot be definitively separated except, of course, intellectually, which amounts to reducing things to the objective. There is no objectivity without subjectivity, and vice versa. The inner world is ambiguous because it involves the physical brain activity of neurons and synapses. The outer world is ambiguous because it contains a residual awareness. The latter is a more controversial statement than the former. It includes the idea that intelligence and creativity are built into the fabric of the cosmos; that they do not merely arise subsequent to the appearance of mankind. An intelligent universe is necessarily ambiguous in just the same way as "creative science" is built upon ambiguity.

Clarity arises from ambiguity in acts of creativity. Ideas structure situations, and situations that are structured in this way appear to be clear and coherent. In the release and euphoria that accompanies the revelation of the deeper structure of a given situation, there is a temptation to believe that a definitive resolution has been attained, that the structure has an objective validity. However, the fluidity of the situation is never lost precisely because the unambiguous retains an element of ambiguity.

I mentioned earlier that the fundamental ambiguity has echoes in two other ambiguities I have mentioned. Take change and stasis and their relationship to the dynamism that is Elliot's still point. You could say, "Fundamentally, things are dynamic...." Or you could look at the fundamental ambiguity as subjectivity/objectivity where the fundamental ambiguity would be called "awareness" and where both subjectivity and objectivity would be two forms of this basic awareness. Finally, the fundamental ambiguity could be called "unity" or unity/duality, because as we saw in chapter 8, unity is ambiguous. It could also be called one/two, which is perhaps the reason why the Pythagoreans did not think of one and two as numbers but as generators of both the mathematical and non-mathematical worlds, of the natural world and of the mind.

The Fundamental Ambiguity as a Description of Science

The first part of this chapter proposed that the fundamental ambiguity offered an alternative way of looking at the world and our place in it. It places ambiguity at the center of the description, with non-ambiguity being relegated to merely one aspect of ambiguity. We saw that this kind of description had the advantage of opening the door to a more open-ended and fluid description of reality that is more process-oriented than object-oriented, a description that is dynamic rather than static.

We can apply this kind of description to science as a whole, but also to particular areas of science. Let me begin with a few words about evolution. We all accept that the natural world at every level, from the astronomical to the cellular, is subject to the process of evolution. Yet we have difficulty accepting that we ourselves—our bodies, our minds, our theories and ideas—also are evolving. Everything is changing, and in our mental life we are in a constant process of falling behind and attempting to catch up. As the Danish science writer Tor Nørretranders[10] has pointed out in his interesting book, *The User Illusion*, there are intrinsic limitations to consciousness. We are doomed to continually play catch-up with the universe. Even when we talk about evolution, we choose to forget about the subject who is doing the talking, about the nature of the device—the mind—that is doing the thinking and theorizing. We do this to avoid our own human situation. If we wish to pursue a science that is based on the total situation, that is holistic and not partial and incomplete, we shall have to face up to the dilemma of what it means to be human.

All of science consists of fundamental dualities. These include the experimental and the theoretical, content and process, science and scientist. There is also the gap between science, per se, and the context that makes science possible: the various social, cultural, and political institutions that support scientific activity. This context, the larger society, not only makes science possible by funding research and paying the salaries of scientists but is itself affected by scientific discoveries and their inevitable technological spin-offs. Conceiving of these dichotomies as ambiguities highlights the fact that there is a tension that must inevitably arise between these dual frames of reference. For example, a good part of this book discussed the difficulties inherent in taking seriously *within a discussion of science* the fact that the scientist is not some disembodied intelligence but a human being fully implicated in every way in her scientific work. Taking the larger picture and not settling for anything that is incomplete and partial—even if this means accepting uncertainty, inconsistency, and conflict—is what is meant by saying that science is ambiguous.

If we take "science as ambiguous" as our starting point, then what we normally think of as science becomes the "face" of the fundamental ambiguity that denies ambiguity. This is the face of clarity, but what is important is that this face is not totally non-ambiguous. Eliminating ambiguity is a cultural and psychological project that is rooted in a desire to escape the human condition. One way of looking at science, the science of certainty, is as an ambitious and far-reaching attempt to eliminate the ambiguous. Much of so-called organized religion is also such an attempt. This does not mean that there is not much in science and religion that is valuable and important, but it does explain the inevitable noxious side effects of both science and religion. Leonard Cohen said it so well:

> And the dealer wants you thinking
> That it's either black or white,
> Thank God it's not so simple
> In my Secret Life.[11]

To reduce things to black or white is the goal of the science of certainty. Cohen's "secret life" cannot possibly be devoid of ambiguity, and neither can science. Our greatest successes in science should not be construed as a rigid set of laws that the universe is obliged to follow, but as a series of patterns, of insights into the complex relationship between certain phenomenal aspects of the natural world that arise *when we look at things from a certain point of view.* Science is a point of view, and there is always room for other points of view. Nevertheless, science approaches something that is universal when it insists on the objectivity of its results. However, this is only so if we recall that the meaning of "objectivity" is itself ambiguous and that we do not mean that the objective eschews the mind but that it avoids that which is colored by personal belief and prejudice. The clarity of science has room within it for the ambiguous and goes seriously astray when this ambiguity is unacknowledged.

CONCLUSION

The "fundamental ambiguity" is a point of view that applies equally well at many different levels of generality. It provides an orientation to science in general, to particular scientific disciplines, and even to subdisciplines within scientific domains.

One could say that "mathematics is ambiguous" if we mean by this that the deeper structure of mathematics is captured by the ambiguous structure/process of the fundamental ambiguity. What we usually call mathematics—the results, proofs, and structure—is the unambiguous face of the subject. The conventional philosophies of math, including formalism and Platonism, emphasize this dimension. Normally, when we ask the question, "What is mathematics?" we are *assuming* that mathematics is an object and are merely questioning what kind of an object it is. This way of asking the question means that we are inevitably blind to the ambiguity in the situation, and as a result we will never succeed in resolving the basic questions about the nature of mathematics—questions like, "Why does mathematics work so well in describing the natural world?" or "Is mathematics discovered or invented?"

Looking at mathematics as a human activity, as mind-dependent, forces one to confront the ambiguous dimension of math. This kind of approach to mathematics arises whenever one considers problems of learning and understanding, the problems associated with doing mathematics. This is a process approach. All of this is in accord with what I called the fundamental ambiguity, which can now be seen as applying to the particular discipline of mathematics.

Within mathematics, the same viewpoint applies to particular parts of the subject, especially those areas that might be consolidated around some proto-concept. I have discussed "number" in chapter 9, so I'll now say a few more words about randomness. One could say that randomness is fundamentally ambiguous and cannot be defined in principle. This is the stance taken by Gregory Chaitin, and before him, Emile Borel. But then we have the face that says randomness is unambiguous, that it comes into being by giving randomness an explicit definition that can be worked with and that has explicit properties, around which specific theories can be developed. But even the unambiguous dimension has room inside it for further ambiguity, as when Chaitin uses the idea of algorithmic randomness to develop the stunning but ambiguous insight that he called Ω.

What is being described here is a fractal structure. We have the following hierarchy: science, mathematics, randomness, and so on. At each level, the statement of the fundamental ambiguity gives us an insight into what is going on; at every level, the same fundamental dynamic of ambiguity plays itself out. Each level is isomorphic in this way to the other levels. Science at all levels possesses a dynamic fractal structure.

Now it could be argued that my ambiguous description of science is really nothing but a duality or complementarity. Two contexts operate in science—the context of discovery and the context of justification or content.[12] One can accept the validity of the distinction between process and content, but still attempt to keep them logically separate. From this point of view, what I am attempting in this chapter could be rejected as merely fuzzy—that is, nonlogical—thinking. I understand and sympathize with this reaction because I have it myself. Ambiguity is just too difficult to contemplate comfortably because there is no secure place to stand and look at the subject spread out before the mind's eye. Logic is that secure point of view and we have an emotional attachment to it (if that is not, in itself, a contradiction) for that very reason. We want to say, "If you are interested in discovery, then discuss discovery and not content. Then, you are free to discuss ambiguity. But if you are discussing content, if you are discussing the timeless truths of mathematics, then stay there where ambiguity has no place."

Ultimately, this book is my answer to this objection. Let me just say that it is my feeling that the world does not divide itself up so neatly into airtight and logically disjointed compartments. You do this when it works and for as long as you can get away with it. It is a first-order approximation. The underlying situation is both more complex and more unified. We confront it for different reasons. The first is theoretical—we get a more accurate and complete picture of science by discarding the artificial separation between process and content. This is especially relevant to the kind of perspective that is called for when thinking about those scientific disciplines that deal explicitly with human beings. The second consideration is that the problems we face as a global society stem to a great extent from this very attempt to divide up reality (and ourselves). The results of science and the critical problems that we face demand that we face up to uncertainty and ambiguity, no matter how stressful this is.

10

∾

Conclusion: Living in a World of Uncertainty

True thinking is thinking that looks disorder and
uncertainty straight in the face.
—Edgar Morin[1]

One of the people who read a draft of an earlier version of this
manuscript asked me how the writing of this book had changed my
own view of science. The short answer is that my view of science has become more complex and, as a result, science is now even more interesting
and exciting to me. I now see the scientific enterprise as a possible way
to live with the total human situation; a vision in which science itself is
an unfolding of the creative potentials of the universe. For a good deal of
my life I had a different vision; one that was entranced by the power of
logical thought and the potential of a grand theory that would explain
everything. I've alluded to these differing visions at different places in
the book. I still love the idea of the grand theory that explains everything, but I now see that I am part of that everything, and that "I" at the
most profound level cannot be explained.

The complexity of science is brought out forcefully in the work of
Edgar Morin, a French sociologist and philosopher who has written extensively about science. His ideas have only recently become available in
English translation,[2] but he is extremely well known in Europe and Latin
America. Morin has not only developed a new way of looking at science,
but he argues that it is vital for the very survival of the human race that
we learn to think in a different way.

The key idea for Morin is complexity both in the natural world and
in the world of thought. Many of the characteristics of what he calls

"complex thinking" will, by now, be quite familiar to readers of this book. For example, complexity necessarily contains elements of uncertainty. He says, "Complexity is not only quantities of units and interactions that defy our possibilities of calculation; it also is made up of uncertainty, indetermination, and random phenomenon." He goes on to say that, "an inescapable imprecision must be accepted not only in phenomena but also in concepts."[3] "Imprecision in concepts" brings to mind my discussion of definition in chapter 1 and proto-concepts in chapter 7.

Complex thought, for Morin, does not follow the dictates of classical logic, but instead follows what he calls "dialogic," which contains both "complementarity and antagonism,"[4] and thus is remarkably similar to Low's (and this book's) use of the term "ambiguity." In fact, the discussion of the fundamental ambiguity in chapter 9 could be read as an exercise in dialogic or complex thinking. This kind of thinking has room for the constructive use of the problematic—inconsistency, paradox, and so on.

Complexity for Morin also emphasizes the "inseparable link between the observer and the observed."[5] This inevitably leads him to think about self-referential systems. Since, as I have repeatedly pointed out, self-reference is inevitably present in any situation in which human beings play a role, it follows that any realistic description of the world must be complex in Morin's sense of the term. Science must become "complex" science, as opposed to what I called classical science in chapter 6. Actually, science *is* complex and the kind of thinking that is required to do science is indeed complex thinking. However, this aspect of science is often unacknowledged when science is reduced to a formal, objective, and static body of immutable truths. The pivotal idea for Morin, and for me, is that it is impossible to definitively separate the objective from the subjective; they are joined in a unity whose complexity arises from the inevitability of self-reference.

SELF-REFERENCE

Self-reference is thus a major factor in any realistic account of science. In a totally different context, let's consider the films of Charlie Kaufman, the writer and film director. Kaufman's work is totally taken up with the implications of self-reference. His most recent film, *Synecdoche, New York,* is both brilliant and disturbing precisely because its theme is self-reference and human life. In a review in *The New York Times,* Manohla Dargis[6] called the film "one of the best films of the year," but added that designating it as such was only "a pathetic response to it's soaring ambition." The main character of the film, a stand-in for Kaufman himself one imagines, gets a MacArthur "genius" grant which he uses to create a play that attempts to be completely true to the director's life. This necessitates making his friends and lovers into characters in the play and eventually leads him to hire an actor to play himself. Of course, the logic of the situation will then force him to find an actor who will play the actor who portrays him and so begins the uncomfortable regression that characterizes self-reference. It all seems a little bizarre, but maybe that is the point.

Normally, we think of movies as telling a story and we observe that story from our seats in the movie theater or living room. We are involved but still safe since we retain a comfortable distance from the situation. Kaufman's movies are uncomfortable because he mixes observer and participant roles and evokes the confusion and complexity that such a mixing generates. As such, he is telling us something very profound about the nature of art, and therefore the nature of life. For doesn't every work of art implicitly mix the observer and the participant, the objective and the subjective? This is the reason that art is important and can touch something that is vital, but this is also the reason that art can be troubling. Art evokes the dilemma of being human, and great art seems to provide a resolution to the dilemma, a transcendence that is an opening to unity. But you cannot have the transcendence without paying the price, which is an authentic confrontation with the ambiguity of existence.

The stakes are precisely the same in science. Great science also needs to be true to the human condition, and to do so it must acknowledge the dilemma of human existence, the dilemma of self-reference implicit in the scientist who tries to understand herself. Classical science positions itself as though it were sitting in an armchair looking out at nature, and this is the posture that is often picked up by society when it makes science into a closed and rigid system of thought. In the ideology of science, its self-referential dimension is nowhere to be seen. If we are prepared to venture into dangerous waters and admit the complexity inherent in trying to make an accurate account of reality, if we are prepared to admit to self-reference, then things get uncomfortable and we must honestly acknowledge and deal with this feeling of discomfort. What do we gain by entering into this difficult domain? What we gain is simple: We gain access to what is real. We would then be not merely avoiding what is uncomfortable. The solution to our problems can evidently only come from the direction of what is real, not from the direction of wishful thinking.

Self-reference in the form of self-consciousness, the ability that we have to stand back and look at ourselves from the outside, is *the* quality that defines what it means to be human. Therefore, self-reference is not just another abstract concept but the essential source of the exquisite agony of being alive. As science aspires to describe the human condition in a manner that is accurate and faithful, this description will have to include, will actually have to originate with, the phenomenon of self-reference. Self-consciousness, and the split in human awareness that makes it possible, produces, as we saw earlier, an infinite cascade of awareness. Not only am I aware of myself, but I am also aware of being aware, and so on. Without acknowledging self-reference in a description of science, we have omitted *the* essential ingredient that brings vitality to the situation.

Whether we are engaged in a discussion of science in general or of some particular problem, like the economic situation, the conversation should start by honestly acknowledging that we are not disinterested observers, we are participants in the situation. We care about these questions. They evidently are important *to us* or why would we be wasting our time considering them? We need to face up to our involvement and not deny it by postulating some kind of pseudo-objectivity. We need to stop using a simplistic vision of science as an excuse for not engaging with the true complexity of the situation. This complexity derives directly from its self-referential nature. We are human beings trying to understand human beings as they try to cope with the world around them, a world that is increasingly dominated by human activity. The first question that must be addressed in any discussion of science is whether or not this element of self-reference is acknowledged. It is a primary feature of my description of science.

Unfortunately, if self-reference is not ignored, the ensuing discussion gets much more complex. Mathematics has taught us that self-reference begets paradox. Therefore, a science that includes self-reference cannot be captured in a theory that is complete and logically airtight (this is why we will never get a mathematically rigorous theory of consciousness). Also, since the self-reference in question involves the very consciousness of the scientist, there is no way to exclude the subjective. This may seem like a very abstract point when it comes to discussing the philosophy of science, but its centrality becomes obvious when considering something like the financial crisis. In this regard, Prof. Jerry Muller said in a discussion of the epistemological roots of our current economic inversion, "What seems most novel [about the financial crisis] is the role of opacity and pseudo-objectivity."[7] The fact that many traders were enriching themselves through their "objective" algorithms and formulas was not an accident but a crucial element of the situation. Ignoring the human side, assuming that the market could regulate itself, was an obvious error in retrospect. Nor can subjectivity be dealt with by the common assumption that everyone involved in the stock market is a "rational actor" who pursues her own economic self-interest. Rationality is another casualty of this crisis; in fact, rationality is often the first thing to go in a crisis. That is how you know it *is* a crisis! The stock market has a lot to teach us, precisely because it shows us so clearly that the assumptions of absolute rationality and objectivity are fallacious.

Now when one is modeling some situation, be it the stock market or a situation in science, it is reasonable to use any assumptions that work, but it is not reasonable to make these assumptions into "laws," or to forget that these are assumptions that people made in the first place. Assumptions are assumptions, even assumptions of objectivity and rationality. Science is too serious an activity, its consequences much too important, for us to leave aside its metaphysical and epistemological assumptions.

THE BLIND SPOT

This book has argued for the existence of a blind spot that needs to be taken into account in our scientific reconstruction of the world. The blind spot exists whether we acknowledge it or not, so the crucial question concerns the nature of our response. Admitting its existence means opening ourselves up to a dynamic source of creativity, but the price of admission involves learning to live with the uncertain. Uncertainty is real. It will emerge whether or not it is acknowledged. But without acknowledgment it cannot be worked with; without working with it there can be no freedom or creativity.

The other choice is denial, which is a natural reaction to something that is unsettling. The essential incompleteness of any system of thought, of one's way of seeing the world, can be very disturbing. The reason for this is our need for control, which is another way of looking at our need for certainty. Certain systems of thought provide the illusion of control. To control something you must be on the outside, an observer, but being on the outside is just one point of view. The problem is that clinging blindly to our certainties, investing all of our energies into the need for control, can have the perverse effect of decreasing our security and increasing the possibility of toxic side effects. Technology is a good case in point—it is wonderful; enriching and exciting—but technology always has consequences that are not foreseen. For example, technological advances may lead to population growth and environmental degradation. Technology is the means through which the project of controlling the human environment is implemented. The "dream" of technology is a dream of total control. But this dream needs to be put aside as the immature fantasy that it is. Control is an illusion; absolute control, even if it were possible, would be a disaster.

The acceptance of intrinsic limits to systems of thought requires a certain humility in our attitude toward the models we use and the theories to which we subscribe. We cannot claim too much for any system. We must always be aware of its limitations, of its optimal range of applicability. In the financial world, this leads us to be cautious with respect to the latest "infallible" mathematical model. In religion, it leads us to value every religion for the insights that it contains but not to accept universal claims to absolute truth. In fact, we would be well advised not to accept any such claims whether they are made by religion or by science. Science and religion must come to see themselves as allies in the search for a human yet transcendent truth. Wonder is just another word for the blind spot—that which is beyond any system.

These comments have an implication for our response to the environmental crisis. The humility I spoke of earlier should make us cautious when we evaluate new ideas and new technologies. If we must err (and we shall inevitably err) we must do so on the side of caution. We cannot, for example, demand rigorous proof of global warming. We cannot demand certainty. There is no such thing as proof outside of deductive mathematical theories. We must ask whether there is a strong likelihood that global warming exists and that human activity is contributing to it. If the answer is yes, then the prudent thing to do is to scale back on those human activities that might contribute to this potential calamity. Let us not take the chance of making a difficult situation even worse.

Another inference that could be drawn from this book is that human judgment must be reclaimed from those theories, ideologies, and mechanical systems that can be so intimidating. Complexity cannot be an excuse for apathy. The problems of the world are the responsibility of all human beings and that responsibility cannot be delegated to some impersonal agency, be it a scientific theory or a super-computer. Human beings must have the confidence to make human decisions about human problems. The world is uncertain and the problems are difficult, so despite our best efforts we will be wrong some of the time. However, if we refuse to engage with problems by putting our faith in some "expertise" that we tell ourselves is beyond our powers to comprehend, we will only contribute to making these problems worse.

The world of the uncertain is the world of creative possibilities. It is the world of freedom, the world of wonder. It is every person's birthright and our collective inheritance. Only our fears and insecurities might lead us to accept less.

Acknowledgments

As the text makes clear, I have found the writing of Albert Low to be an unending source of inspiration. Of course, he is not responsible for any specific views that I put forward in this book. It is his unique perspective that I value. I'd also like to thank Joseph Auslander, Gregory Chaitin, Jean Nantel, Ron Rower, and Michael Schleifer for their input and support. Vickie Kearn has been an invaluable source of encouragement, practical advice, and feedback at every stage of this project. Finally, thanks to Miriam for listening, and for everything else.

Notes

Preface. The Revelation of Uncertainty

1. Friedman, Thomas L., "Elvis Has Left the Mountain," op-ed column, *The New York Times*, Feb. 1, 2009.

Chapter 1. The Blind Spot

1. Goldstein, (2005), p. 191.
2. Einstein, (1930), pp. 193–94.
3. http://en.wikipedia.org/wiki/Blind_spot_%28vision%29.
4. http://plato.stanford.edu/entries/spacetime-singularities/.
5. Kauffman, (2008), pp. 5–6.
6. Pirsig, (1974), pp. 205–6.
7. Mathematicians have defined infinity in various ways, as I discuss in chapters 3 and 4 of *How Mathematicians Think*.
8. Moore, (1990).
9. Chaitin, (2005), p. 124.
10. Greenberg, (1974), pp. 9–10.
11. In chapter 8, I shall call number a "proto-concept" to emphasize its open-ended, generative possibilities.
12. Schwartz and Begley, (2002), pp. 65–66.
13. Of course, it is sometimes possible for gut feelings to lead you astray. However, it is equally misleading to assert that the only way to validate a gut feeling is by translating it into something else entirely.
14. Heschel, (1951), p. 4.
15. Byers, (2007), p. 304.

Chapter 2. The Blind Spot Revealed

1. Chaitin, (2007), p. 333.
2. Ibid.
3. As I write this book, I am aware that it is like a dialogue between two parts of myself—between my own need for certainty, on the one hand, and my need for

wonder, on the other. I've been told that I write in two different styles, one is dry and rational, even pedantic; the other is more lyrical. In fact, one motivation for writing a book is to unify these two pieces of myself in exactly the same way as I am trying to point to a possible unification of the diverse tendencies within science.

4. Tipler, (1994).

5. Zajonc interviewed on the CBC radio program *Ideas*.

6. Kline, (1980).

7. "Equal" in this context means that all elements of the first set can be matched (i.e., put into a one-to-one correspondence) with all elements of the second. This paradox is explained by noting that two sets that are equal in this sense (they have the same cardinality) do not necessarily have the same elements and so are not equal as sets.

8. Cf. p. 31.

9. This follows from the work of Kurt Gödel and Paul Cohen, who together showed that the continuum hypothesis cannot be either proven or disproven on the basis of the standard Zermelo–Fraenkel axioms for set theory.

10. By making mathematics into a "game" with no meaning other than logical consistency, formalism in mathematics has shielded mathematicians from uncertainty. But the price has been that a great deal of its meaning and significance has been lost.

11. Aczel, (2000), p. 132.

12. They did not succeed in doing what they set out to do—namely, reducing mathematics to logic. In fact, it was a glorious failure, but their attempt was valuable and important and set the stage for the work of Gödel.

13. This is the essence of Gödel's results and of his method. He embraces paradox and self-reference but emerges with concrete results.

14. Smale, (2000), pp. 271–94.

15. Quotations in this section are taken from Goldstein, (2005), pp. 99–107.

16. Chapters 3 and 6.

Chapter 3. Certainty or Wonder?

1. Herschel, (1830), *A Preliminary Discourse on the Study of Natural Philosophy*, as quoted in Holmes, (2009), p. xiii.

2. Holmes, p. xvi.

3. Dawkins, (1998), p. 27.

4. In books such as *The God Delusion* (2006) and *The Blind Watchmaker* (1986).

5. Einstein, (1930), pp. 193–94.

6. Greene, (2000).

7. The connection between science and scientist, between subject and object, will be taken up in chapter 6. It is something that is discussed by the French sociologist and philosopher of science, Edgar Morin, whose work I discuss in chapter 10.

8. Michael Ignatieff, *The Needs of Strangers,* quoted in Thien (2007).

9. Russell, *Portraits from Memory,* quoted in Davis and Hersh (1981), p. 333.

10. I do not say that working scientists have such a view, so much as people believe that this is what it means for something to be scientific. In the popular mind, science is identified with certainty.

11. Hersh and John-Steiner, (2010).

12. Winer, (1994), p. 93.

13. Albers and Alexanderson, (1985), p. 127.

14. "The Elegant Universe," *Nova,* PBS, or the book by Brian Greene on which it is based.

15. Abram, (1996), p. 34.

16. Cf. the discussion on pp. 167–68.

Chapter 4. A World in Crisis!

1. Interviewed on PBS's *Charlie Rose* program, quoted in Dennis Overbye, "They Tried to Outsmart Wall Street," *The New York Times,* March 10, 2009.

2. Taleb, (2004).

3. Taleb, (2007).

4. cf. Dennis Overbye, *They Tried to Outsmart Wall Street,* or the book by Emanuel Derman (2004).

5. The Black–Scholes model "had figured out how to price and hedge these options in a way that seemed to guarantee profits." The model was created by Fisher Black, then at University of Chicago, and Myron S. Scholes and Robert C. Merton, both at MIT (taken from the Overbye article).

6. Porter, (1995), p. 5–6.

7. Quotations in this paragraph are taken from Taleb (2008).

8. cf. the discussion of Kuhn in chapters 5 and 6.

Chapter 5. Ambiguity

1. Zajonc, (2006), p. 3.

2. Suri and Bal, (2007).

3. Ibid., p. 245.

4. Compare Low, (2002), pp. 20–21.

5. Quoted in Fitzgerald and James, (2007), p. 59.

6. Bernstein, (1976), pp. 39–41.

7. Empson, (1966).

8. Koestler, (1964).

9. Quotation from the article "Charting Creativity: Signposts of a Hazy Territory" by Patricia Cohen, *The New York Times,* May 7, 2010.

10. Low, (1993), p. 50.

11. Another famous gestalt picture, two faces and/or a vase, was used as the dust jacket picture for my book *How Mathematicians Think*.

12. Kuhn, (1962), pp.150–51.

13. Cf. Gray and Tall, (1994).

14. Thurston, (1990), p. 846.

15. Gray and Tall.

16. It should however be noted that integration is more general—that is, all continuous functions are integrable, but "most" continuous functions are nowhere differentiable.

17. Fermat's Theorem states that for $n > 2$, the equation $x^n + y^n = z^n$ has no integer solutions. See for comparison the discussion on pages 168–69.

18. Mazur, (2008), p.1.

19. Herbert, (1985), pp. 66–67.

20. Bodanis, (2000).

21. I define what I mean by "classical science" in chapter 6. Basically, it means the absence of self-reference, putting aside the reality that science is created by human beings and therefore influenced by their participation in the situation being described.

22. Dyson, Oct. 25, 2007, pp. 45–47.

23. Chaitin, "Less Proof, More Truth, Review of *How Mathematicians Think*," (2007).

24. *The New York Times*, Nov. 6, 2008.

Chapter 6. Self-Reference: The Human Element in Science

1. *New Oxford American Dictionary*.

2. Ibid.

3. Lakatos, (1976).

4. Hersh, (1997).

5. Albert Low, *On Two Ambiguities*, private communication.

6. Dunham, (1990).

7. It's a little like reading or writing a critical analysis of a movie or a piece of music that you love. It enriches the experience in one way but in another you lose the feeling of immediacy that comes with the pure unmediated experience of watching or listening.

8. Cf. Low, (2002), pp. 79–82, 97–99.

9. Feynman, (1985), pp. 9–10.

10. Herbert, (1985), pp. 66–67.

11. Kuhn, (1962).

12. Ibid., p. 151.

13. Chaitin, "Less Proof, More Truth, Review of *How Mathematicians Think*," (2007).

Chapter 7. The Mystery of Number

1. Davis, (1965).

2. Cf. diagram in chapter 4.

3. *New Oxford American Dictionary*.

4. Shanks, (2001).

5. Complex numbers are numbers like $2 + 3i$—for example, where $i^2 = -1$.

6. Livio, (2002).

7. Davis and Hersh, (1981), pg. 97.

8. $1729 = 1^3 + 12^3 = 9^3 + 10^3$.

9. Tammet, (2006), pp. 3–7.

10. Cf. Luria, (1968).

11. Sacks (1985), chapter 23, *The Twins*.

12. Ibid., p. 207.

13. Dehaene, (1997).

14. Cf. Livio, (2002).

15. Cf. Maor, (1994).

16. Cf. Nahin, (1998).

17. For example, do each of the digits 0, 1, 2, 3, ... 9 occur equally often in its decimal representation.

18. For example, Taylor (2009).

19. Davis, Philip J., *When Is a Problem Solved?* in Gold and Simons (2008), p. 83.

Chapter 8. Science as the Ambiguous Search for Unity

1. John Lennon, *Imagine*, Lennon Music, (1971).

2. Maxwell and Tschudin, (1990), p. 47.

3. O'Brien, (1964), pp. 80–81. Quoted in A. Low (1993).

4. *New Oxford American Dictionary*.

5. This is the statement that the equation $x^n + y^n = z^n$ has no integer solutions for $n > 2$. It was proved through the efforts of a whole series of mathematicians culminating in the work of Andrew Wiles in 1993.

6. This states that *every* even number is the sum of two prime numbers (e.g., $10 = 7 + 3$ and $50 = 43 + 7$).

7. These qualitative properties of the natural numbers have regularly been rediscovered. When Jungian psychologists, for example, talk about numbers as arche-

types, which have standing both in the natural and mental worlds (cf. Von Franz), they are referring to the qualitative properties of numbers.

8. Wigner, (1960).

9. Cohen, (2007), p. 1.

10. Ibid. The mathematicians John Couch Adams and Urbain le Verrier.

11. Barrow, (1992), p. 1.

12. Cf. Lakoff and Núnez (2000) and Lakoff and Johnson (1980).

13. Hersh, (1997).

14. Doidge, (2007), Schwartz and Begley (2002), and Siegel (2007).

15. Pp. 81–82.

16. This conjecture has now been proved (by Christophe Breuil, Brian Conrad, Fred Diamond, and Richard Taylorand). It is sometimes called the modularity theorem, c.f. Darmon (1999).

17. The Rosetta stone was a kind of dictionary that enabled archeologists to decipher Egyptian hieroglyphics. It contained Egyptian demotic, ancient Greek, and hieroglyphics, the first two of which were already understood.

18. Quoted in Singh, (1997).

19. Pirsig, (1974), p. 164.

20. *Stanford Encyclopedia of Philosophy*, http://plato.stanford.edu/entries/paradox-zeno/#Zeninf.

21. *New Oxford American Dictionary*.

22. *Webster's Ninth New Collegiate Dictionary*.

23. *Encarta World English Dictionary*.

24. Lakoff and Johnson, (1980).

25. The numbers π (the ratio between the circumference and diameter of a circle) and e (the base of the natural logarithms) are known to be transcendental, but whether sums and products of these two numbers, like $\pi + e$, $\pi - e$, $\pi \cdot e$, and π/e, are transcendental is unknown.

26. Chaitin, (2007), *How Real Are the Real Numbers?*, pp. 267–80.

27. Blum, Cucker, Shub, and Smale, (1997). Introduction.

28. Ibid.

29. Actually, quantum mechanics can be formalized in two ways: one that is discrete; another that is continuous.

30. Becker, (1973).

31. Cf. Becker, pp. 68–70.

32. Winer, (1994), p. 93.

33. Taylor, (2009), p. 49.

34. Lightman, (2005), p. 16.

35. Cf. Byers, pp. 328–30.

36. The poem is by Matsuo Basho. The book on his poetry by Aitken (1978) is highly recommended. Chapter 1, pp. 25–29, consists of an analysis and commentary on this haiku.

CHAPTER 9. THE STILL POINT

1. Eliot, (1999), p. 5.

2. Actually, his statements were much more subtle than that. He made the enigmatic statement, "*We both step and do not step in the same rivers. We are and are not,*" which not only challenge the idea of the world as being made up of objects, but also the logical rules (like the law of identity) that are used to analyze the world.

3. The idea that science will one day conquer death is something that comes up with a certain regularity. For example, a book by Frank Tipler called *The Physics of Immortality* focused on this. A number of scientists look on aging as a disease and believe that the human life span can be extended a great deal. If perhaps enhanced with technological advances, such as artificial body parts, human beings could aspire to live forever. The idea of immortality hovers in the background of the scientific enterprise and tells us something about its nature. This is another of the conventional roles of religion that has been taken over by science.

4. Low, (2002).

5. The next few paragraphs essentially reproduce the substance of a note I sent to Low after reading his statement, which he reproduced in his book, *Creating Consciousness*, pp. 259–60.

6. It is interesting to compare the statement of the fundamental ambiguity which says of itself that it cannot be understood, with the proposition that arises in the proof of Gödel's incompleteness theorem, which revolves around producing a mathematical proposition that says of itself that it cannot be proved. cf. Byers, p. 273.

7. cf. Block, Guckenheimer, Misiurewicz, and Young.

8. Bohm, (1980), p. 4.

9. Remember that when I discuss objectivity and subjectivity here, I am not concerned with whether or not the given topic is subject to idiosyncratic feelings or opinions.

10. Nørretranders, (1998). You might consider his book to be an argument for the existence of a blind spot that is built into human consciousness.

11. Leonard Cohen, "In My Secret Life," *Ten New Songs*, (2001).

12. This distinction is due to Popper, and was pointed out to me by my old friend, the late philosopher Gerald A. Cohen.

CHAPTER 10. CONCLUSION: LIVING IN A WORLD OF UNCERTAINTY

1. Morin, (2008), p. 88.

2. Ibid.

3. Ibid., pp. 20–21.

4. Ibid., p. 49.

5. Ibid., p. 4.

6. Dargis, Manohla, "Dreamer, Live in the Here and Now," *The New York Times*, Oct. 24, 2008.

7. Quoted in the column, "Greed and Stupidity," by David Brooks in *The New York Times*, op-ed column, April 3, 2009.

References

Abram, David, (1996). *The Spell of the Sensuous: Perception and Language in a More-than-Human World*, Vintage Books, New York.

Aczel, Amir D., (2000). *The Mystery of the Aleph: Mathematics, the Kabbalah, and the Search for Infinity*, Pocket Books, Simon & Schuster, New York.

Aitken, Robert, (1978). *A Zen Wave: Basho's Haiku and Zen*, Weatherhill, New York.

Albers, D. J., and Alexanderson, G. L., (Eds.), (1985). *Mathematical People: Profiles and Interviews*, Birkhäuser, Boston.

Barrow, John D., (1992). *Pi in the Sky: Counting, Thinking, and Being*, Clarendon Press, Oxford.

Becker, Ernest, (1973). *The Denial of Death*, The Free Press, New York.

Bernstein, Leonard, (1976). *The Unanswered Question: Six Talks at Harvard*, Harvard University Press.

Block, L., Guckenheimer, J., Misiurewicz, M., and Young, L.-S., (1980). *Periodic Points and Topological Entropy of One Dimensional Maps*, Lect. Notes Math., No. 819, Springer, New York, pp. 18–34.

Blum, Lenore, Cucker, Felipe, Shub, Michael, and Smale, Steve, (1997). *Complexity and Real Computation*, Springer, New York.

Bodanis, David, (2000). $E = mc^2$: *A Biography of the World's Most Famous Equation*, Anchor Books, Random House, New York.

Bohm, David, (1980). *Wholeness and the Implicate Order*, Routledge, London.

Byers, William, (2007). *How Mathematicians Think: Using Ambiguity, Contradiction, and Paradox to Create Mathematics*, Princeton University Press, Newark, NJ.

Chaitin, Gregory, (2005). *Meta Math! The Quest for Omega*, Vintage Books, Random House, New York.

———, (2007). *Thinking about Gödel and Turing: Essays on Complexity, 1970–2007*, World Scientific Publishing, Hackensack, NJ.

———, "Less Proof, More Truth, Review of *How Mathematicians Think*," *New Scientist*, July 28, 2007.

Cohen, Daniel J., (2007). *Equations from God*, Johns Hopkins University Press, Baltimore.

Cohen, Leonard, (2001). "In My Secret Life," *Ten New Songs*, Sony Music.

Darmon, Henri, (1999). *A Proof of the Full Shimura-Taniyama-Weil Conjecture is Announced*. Notices of the A. M. S., December, 1397–401.

Davis, Philip, (1965). *The Mathematics of Matrices: A First Book of Matrix Theory and Linear Algebra*, Blaisdell Pub. Co., New York.

Davis, Philip, and Hersh, Reuben, (1981). *The Mathematical Experience*, Birkhäuser, Boston.

Dawkins, Richard, (1986). *The Blind Watchmaker*, W. W. Norton & Co., New York.

———, (1998). *Science, Delusion, and the Appetite for Wonder*, Houghton Mifflin, Boston.

———, (2006). *The God Delusion*, Houghton Mifflin Harcourt, New York.

Dehaene, Stanislas, (1997). *The Number Sense: How the Mind Creates Mathematics*, Oxford University Press, Oxford, UK.

Derman, Emanuel, (2004). *My Life as a Quant: Reflections on Physics and Finance*, John Wiley & Sons, Hoboken, NJ.

Doidge, Norman, (2007). *The Brain that Changes Itself: Stories of Personal Triumph from the Frontiers of Brain Science*, Penguin Books, New York.

Dunham, W., (1990). *Journey through Genius: The Great Theorems of Mathematics*, John Wiley & Sons, New York.

Dyson, Freeman, "Working for the Revolution," *New York Review of Books*, Vol. 54, No. 16, Oct. 25, 2007.

———, "Our Biotech Future," *New York Review of Books*, Vol. 54, No. 12, July 19, 2007.

Einstein, Albert, "What I Believe," *Forum and Century*, 84, Oct. 1930, pp. 193–94.

Eliot, T. S., (1999 edition). *Four Quartets*, Faber and Faber, London.

Empson, William, (1966). *Seven Types of Ambiguity*, New Directions Publishing Co., New York.

Feynman, Richard, (1985). *QED: The Strange Theory of Light and Matter*, Princeton University Press, Princeton, NJ.

Fitzgerald, Michael, and James, Ioan, (2007). *The Mind of the Mathematician*, The Johns Hopkins University Press, Baltimore.

Gold, Bonnie, and Simons, Roger, (Eds.), (2008). *Proof and Other Dilemmas: Mathematics and Philosophy*, Mathematical Association of America.

Goldenfeld, Nigel, and Woese, Carl. "Biology's Next Revolution," *Nature*, 445, 369, Jan. 25, 2007.

Goldstein, Rebecca, (2005). *Incompleteness: The Proof and Paradox of Kurt Gödel*, W. W. Norton, New York.

Gray, E., and Tall, D., "Duality, Ambiguity, and Flexibility: A 'Proceptual' View of Simple Arithmetic," *Journal for Research in Mathematics Education*, 25, 2, 1994, pp. 116–40.

Greenberg, Marvin Jay, (1974). *Euclidean and Non-Euclidean Geometries: Development and History*, W. H. Freeman and Company, San Francisco.

Greene, Brian, (2000). *The Elegant Universe: Superstrings, Hidden Dimensions, and the Quest for the Ultimate Theory*, Vintage Books, Random House, New York.

Halmos, P. R., (1985). *I Want to Be a Mathematician*, MAA Spectrum, Springer Verlag, New York.

Herbert, Nick, (1985). *Quantum Reality: Beyond the New Physics*, Anchor Books, Doubleday, New York.

Hersh, Reuben, (1997). *What Is Mathematics Really?* Oxford University Press, New York.

Hersh, Reuben, and John-Steiner, Vera, (2010). *Loving and Hating Mathematics: Inside Mathematical Life*, Princeton University Press, Princeton, NJ.

Heschel, Abraham Joshua, (1951). *Man Is Not Alone*, Farrar, Straus, & Giroux, New York.

Holmes, Richard, (2009). *The Age of Wonder: How the Romantic Generation Discovered the Beauty and Terror of Science*, Pantheon Books, New York.

Kapleau, Philip, (1980). *The Three Pillars of Zen*, Anchor Books, New York.

Kauffman, Stuart A., (2008). *Reinventing the Sacred: A New View of Science, Reason, and Religion*, Basic Books.

Kline, Morris, (1980). *Mathematics: The Loss of Certainty*, Oxford University Press, New York.

Koestler, Arthur, (1964). *The Act of Creation*, Picador, Pan Books, London.

Kuhn, Thomas S., (1962). *The Structure of Scientific Revolutions, 2nd ed.*, The University of Chicago Press, Chicago.

Lakatos, I., (1976). *Proofs and Refutations: The Logic of Mathematical Discovery*, Cambridge University Press, New York.

Lakoff, G., and Johnson, M., (1980). *The Metaphors We Live By*, The University of Chicago Press, Chicago.

Lakoff, G., and Núnez, Rafael E., (2000). *Where Mathematics Comes From: How the Embodied Mind Brings Mathematics into Being*, Basic Books, New York.

Lightman, Alan, (2005). *A Sense of the Mysterious: Science and the Human Spirit*, Pantheon Books, New York.

Livio, Mario, (2002). *The Golden Ratio: The Story of Phi, the World's Most Astonishing Number*, Broadway Books, New York.

Low, Albert, (1993). *The Butterfly's Dream: In Search of the Roots of Zen*, Charles E. Tuttle, Boston.

———, (2002). *Creating Consciousness: A Study of Consciousness, Evolution, and Violence*, White Cloud Press, Ashland, OR.

———, (2008). *The Origin of Human Nature: A Zen Buddhist Looks at Evolution*, Sussex Academic Press, Brighton, UK.

Luria, A. L., (1968). *The Mind of a Mnemonist*, Harvard University Press, Cambridge, MA.

Maor, Eli, (1994). *e: The Story of a Number*, Princeton University Press, Princeton, NJ.

Maxwell, Meg, and Tschudin, Verena, (1990). *Seeing the Invisible: Modern Religious and Other Transcendent Experiences*, Arkana, London.

Mazur, Barry, (2008). *When Is One Thing Equal to Some Other Thing?* in Gold and Simons, eds. *Proof and Other Dilemmas*, Mathematical Association of America, Washington, DC.

Moore, A. W., (1990). *The Infinite*, Routledge, London.

Morin, Edgar, (2008). *On Complexity, (Advances in Systems Theory, Complexity, and the Human Sciences)*, Hampton Press, Cresskill, NJ.

Nahin, Paul J., (1998). *An Imaginary Tale: The Story of $\sqrt{-1}$*, Princeton University Press, Princeton, NJ.

Nørretranders, Tor, (1998). *The User Illusion: Cutting Consciousness Down to Size*, Viking, Penguin, New York.

O'Brien, Elmer, (1964). *The Essential Plotinus*, New American Library, Mentor Books, pp. 80–81. Quoted in A. Low (1993).

Pirsig, Robert M., (1974). *Zen and the Art of Motorcycle Maintenance: An Inquiry into Values*, William Morrow & Company, New York.

Porter, Theodore M., (1995). *Trust in Numbers: The Pursuit of Objectivity in Science and Public Life*, Princeton University Press, Princeton, NJ.

Prigogine, Ilya, (1980). *From Being to Becoming: Time and Complexity in the Physical Sciences*, W. H. Freeman and Co., San Francisco.

———, (1997). *The End of Certainty: Time, Chaos, and the New Laws of Nature*, Free Press, Simon & Shuster, New York.

Sacks, Oliver, (1985). *The Man Who Mistook His Wife for a Hat and Other Clinical Tales*, Harper and Row, New York.

———, (2008). *Musicophilia: Tales of Music and the Brain*, Vintage Books, Random House.

Schwartz, Jeffrey M., and Begley, Sharon, (2002). *The Mind and the Brain: Neuroplasticity and the Power of Mental Force*, Harper Perennial, New York.

Shanks, Daniel, (2001). *Solved and Unsolved Problems in Number Theory*, American Mathematical Society, Chelsea Publishing Company, Providence, RI.

Siegel, Daniel J., (2007). *The Mindful Brain: Reflection and Attunement in the Cultivation of Well-Being*, W. W. Norton, New York.

Singh, Simon, (1997). *Fermat's Enigma: The Epic Quest to Solve the World's Greatest Mathematical Problem*, Penguin Books.

Smale, Steve, (2000). *Mathematical Problems for the Next Century, Mathematics: Frontiers and Perspectives*, pp. 271–94, American Mathematics Society, Providence, RI.

Suri, Gaurav, and Bal, Hartosh Singh, (2007). *A Certain Ambiguity*, Princeton University Press, Princeton, NJ.

Taleb, Nassim Nicholas, (2004). *Fooled by Randomness: The Hidden Role of Chance in Life and in the Markets*, Random House.

———, (2007). *The Black Swan: The Impact of the Highly Improbable*, Random House, New York.

———, *The Fourth Quadrant: A Map of the Limits of Statistics*, Edge: The Third Culture, (Sept. 15, 2008), http://www.edge.org/3rd_culture/taleb08/taleb08_index.html.

Tammet, Daniel, (2006). *Born on a Blue Day*, Hodder & Stoughton, London.

Taylor, Jill Bolte, (2009). *My Stroke of Insight: A Brain Scientist's Personal Journey*, Plume, New York.

Thien, Madeleine, (2007). *Certainty*, Little, Brown and Co., New York.

Thurston, William, "Mathematical Education," *Notices of the American Mathematical Society*, 37, 1990, pp. 844–50.

Tipler, Frank J., (1994). *The Physics of Immortality: Modern Cosmology, God, and the Resurrection of the Dead*, Anchor, Random House, New York.

Von Franz, Marie-Louise, (1974). *Number and Time: Reflections Leading toward a Unification of Depth Psychology and Physics*, Northwestern University Press, Evanston, IL.

Wigner, E., "The Unreasonable Effectiveness of Mathematics in the Natural Sciences," *Comm. Pure & App. Math.*, 13, 1, 1960.

Winer, Robert, (1994). *Close Encounters: A Relational View of the Therapeutic Process*, Jason Aronson Inc., Northvale, NJ.

Woese, Carl R., "A New Biology for a New Century," *Microbiology and Molecular Biology Reviews*, June 2004, pp. 173–86.

Zajonc, Arthur, "Cognitive-Affective Connections in Teaching and Learning," *Journal of Cognitive Affective Learning*, 3(1), 2006, pp. 1–9.

Index